JOURNAL OF A WHALING VOYAGE

KEPT BY A GREEN HORN IN THE FORECASTLE OF THE SHIP NIMROD COMMENCING NOV. 1ST, 1842

EDITED BY JOHN BLACK

Journal of a Whaling Voyage
Kept by a Green Horn in the Forecastle of the Ship Nimrod Commencing Nov. 1st, 1842
All Rights Reserved.
Copyright © 2020 John Black
v5.0

The opinions expressed in this manuscript are solely the opinions of the author and do not represent the opinions or thoughts of the publisher. The author has represented and warranted full ownership and/or legal right to publish all the materials in this book.

This book may not be reproduced, transmitted, or stored in whole or in part by any means, including graphic, electronic, or mechanical without the express written consent of the publisher except in the case of brief quotations embodied in critical articles and reviews.

Outskirts Press, Inc.
http://www.outskirtspress.com

ISBN: 978-1-9772-2326-5

Cover Photo © 2020 Wikimedia Commons. All rights reserved - used with permission.
Other illustrations in the log are from:
Incidents of a Whaling Voyage
By Francis Allyn Olmstead – 1841

Outskirts Press and the "OP" logo are trademarks belonging to Outskirts Press, Inc.

PRINTED IN THE UNITED STATES OF AMERICA

Dedicated to my children Adam Nathaniel Black and Anna Michaela Black who heard this story from their grandmother and to their Great Grandfather Graham Lee who lived this adventure. And to my wife, Susa Morgan Black who put up with me during this process.

"They that go down to the sea in ships,
That do business in great waters;
These see the works of the Lord,
And his wonders in the deep."
Psalm 107:23, 24

Contents

From the Editor: i
Listing of images iii
The Log Book 1
Crew List 135
Graham Lee's shipboard will. 137

Appendix A: Graham Lee Bio. 139
Appendix B: Vessels and Terminology. 141
    Sails 147
    Types of Whaling Ships 149
    Types of whaleship rigs: 151
    The Whaleboat 155
Appendix C: The Varieties of Whales 159
Appendix D 163

From the Editor:

In the following pages will be found a copy of the Log Book kept by Graham Lee, from November 1842 to January 1845, while employed on the whaleship *Nimrod*. In making this printed copy, care has been taken to preserve, so far as possible, the characteristics peculiar to his hand writing, such as punctuation, capitalization, and spelling while making it understandable to the 21st Century reader.

It was found among the papers of my maternal grandmother, Anna Mary Lee Heisey, his youngest of 14 children and only daughter. I had heard about this log book but had never seen it before.

It was found that in a few instances words were, for one reason and another, so obliterated that it was almost impossible to obtain a correct spelling and in such instances, the whole words were copied to conform to the spelling as they seemed to be, regardless of their construction or spelling. Whenever a case of this kind occurs it will be followed immediately by a question (?) mark in parenthesis. It will still be understood that this question mark is not found in the original book but is the insertion of the copyist indicating doubt about the word which precedes it.

Some discrepancies are also found in the dating without any apparent cause. In all such instances a star (*) is used calling attention to an explanatory NOTE below. The star is also not in the original book but is used only as a means of explanation in copy.

When only part of a word is legible the remainder is marked as ??.

Listing of images

Dedications:

Page iii - Graham Lee photo

FRONTPIECE: Perils of Whaling

Page 33 - The Attack

Page 35 - Pulling teeth

Page 95 - Old Native Church Native church

Page 141 - Found in old stack of family photos.

Page 154 - Whaleship Charles W. Morgan

Page 155 - by Francis Allyn Olmstead – 1841 pub. 1841.

Page 158 - Tub, harpoon, lance

Page 159 - float

The Log Book

Tuesday November 1st 1842 –– Left home for New Bedford with Alexander Holley - who accompanied me as far as New York where we arrived on the following morning. Little did I think that leaving home for 3 years was to cause my parents, brothers, sisters & friends so much pain had I anticipated it I never should have parted with them until compelled to do it. I had supposed that besides family friends I had no other; but at parting I discovered my mistake & ere had left my native town I repented my determination and wished myself again with my friends.

Wednesday November 2nd — Spent most of the day in visiting the Groton water Aqueducts with H. N. Baldridge. Left New York at 4 o'clock in the steamboat *Massachusetts* for New Bedford & passed the night in lounging upon deck & conversing with an old sailor.

Thursday November 3rd — Parted with the last of my acquaintances (Mr. T. Chittenden) at Providence this morning. Arrived in New Bedford at 10 AM & immediately started in search of a place. After obtaining the promise of a berth I looked for a boarding place & thanks to the friends of seamen found good one at the Sailors Home kept by Mr. Nathaniel Hathaway, No. 29 Hater St.

From the 4th to the 14th I have spent the time in lounging about town & in watching the completion of my future home upon the wave.

Tuesday November 15th — Arose just in time to see the *Nimrod* moving majestically into the stream with all my traps aboard save some books which a bookseller kindly let me have at half price because my means could go no farther. At 8 o'clock I went on board in a pilot boat with about 20 others who I soon found were to be my shipmates. After working about two hours to cat an anchor the pilot boat came along side for the last time with the Owners, Captain & remainder of the crew. As soon as he came within hailing he sang out to weigh anchor and at it we went. After heaving away for some time the ship began to move slowly to leeward & then as we piled on the canvas she laid her course for the open sea on which she is to rock for many a month & perhaps go to the bottom.

At noon the officers went into the cabin & the hands into the forecastle there to partake of their hardtack & hard enough I found it. Nothing

The Log Book

but salt beef, potatoes & hard sea biscuit brought to us in wooden kids & set on a chest or on the floor (just as it happens more frequently on the latter however).

As soon as it is set down all hands fell at it with a knife in one hand & five fingers in the other - he that gets the most & first is the best fellow.

At two o'clock the pilot left us with the owners and visitors & with them the last link between us & our native land. At sundown we were called aft & divided into two watches. As good luck would have it I am in the Captain's & second mate's watch which is considered preferable to the first mate's watch. The Captain has forbidden all swearing & fighting saying that he would do that himself. The two watches are called the larboard & starboard, the Captain's being the starboard. The day is divided into six watches, one from 12 to 4, from 4 to 7, from 7 to 11 PM, from 11 to 3, from 3 to 7 and from 7 to 12.

Wednesday November 16th –– Today it has stormed most of the time & the ship has been under double reefed topsails. Green hands begin to grow sick, myself among the rest. If anything can make one think of home it is seasickness & nothing but hardtack to eat. How many minds revert to the good things of mother's pantry & how many wish themselves by father's fire instead standing watch in a dark, rainy night. I cannot deny but that I have wished myself during the past day & night snug at home with a plenty to eat & a good bed at night, yet I continue to console myself with a hope that there is a better day coming, for unless I succeed in this business I shall get some experience which may be worth something.

Thursday November 17th –– Weather clear but windy, have had to assist in taking in sail & in reefing the mizzen top sail. I bound the yard to the shrouds & made a nice job for a boat steerer to cast it off. As good luck would have it none of the officers knew who did it.

Friday November 18th — The wind has blown almost a hurricane today. The ship has been under close reefed main top sail, shipping a sea every few minutes keeping us all wet.

Saturday November 19th — Weather pleasanter than yesterday but squally. Ship under easy sail. Affairs are getting rather more pleasant in the forecastle, the sick are getting well & the well better. For several days past the forecastle has been a complete den of filth. The sick vomiting upon chests, clothes, & everything else.

Sunday November 20th — On rising at 7 o'clock I was agreeably surprised by seeing one of the finest mornings I ever beheld, perhaps the stormy weather of late makes the sun look brighter & everything else pleasanter then it otherwise would. At any rate I never enjoyed myself more at home than I did this morning. (I mean as far as the elements are concerned). At 8 o'clock the Captain (who by the bye I like the better the more I see of him) called all hands aft and gave all that had none a Bible & testament, afterwards distributing papers & tracts among them. My self and many others spent the day in reading. The wind being northerly we let the ship run dead ahead of it so that there was but little to be done in the way of handling sail. I have today seen a fair example of the improvidence and carelessness for the morrow which is characteristic of the sailor. Some who have not half a "fit out" are trading things which they absolutely need for mere trifles. One fellow gave another 6 lbs. of tobacco which would have been worth $6.00 to him for a book not worth 50 cents.

We have today had something extra for dinner, what it was I am unable to tell without consulting the cook records. I suppose however it was a sort of pudding made from the crumbs of sea biscuit sweetened with "lasses." Our meals are generally the same, beef, potato & sea biscuit. Coffee for breakfast and tea for supper. The coffee tastes much like hot water seasoned with boneset. The tea is inferior to a decoction of oak leaves. The beef & bread are so tough that my jaws are sore with masticating it. For table we use our chests, for furniture a quart cup, tin pan, spoon & knife. With these we eat & drink as we can & as necessity is the mother of invention we succeed tolerably well, at least a bystander would judge so from the manner in which the hard tack disappears.

Monday November 21st — As I lay dreaming of home & friends in Salisbury, I could imagine myself and a circle of them enjoying all manner of sport. In fact I began dreaming of making a visit to Mount Riga with some others when a shrill voice called out "Starboard watch on deck." Never was a fellow more disappointed than myself for instead of calling on friend Fish, I was called upon to appear on deck to do the bidding of Captain or mates. The day has been very pleasant & the sea smoother than since we left New Bedford. I believe we are in latitude 34º 30' north, sailing southerly with a fair wind.

We have been most of the day hauling out and stowing away things between decks & preparing things for business in case we get a whale alongside. For two days past we have had a lookout at masthead but as yet he has discovered nothing. All I ask for is that we may in two years run into New Bedford with a full cargo. Not because I wish to get out of the business as soon as possible but I wish to relieve my parents of their anxiety for my safety. I should like well to return a year before I am expected and astonish them with my presence. I have had a severe cold today in consequence of keeping watch without a coat last night.

Tuesday Nov. 22nd — Weather very pleasant in the forenoon all sail set. Towards night the wind freshened into a gale. We were obliged to put her under close reefed top sails & fore sail. Spent the day in setting up & taring rigging. Took the wheel for the first time today. Found it is no child's play to keep the ship in her place --- --- (?). We are now sailing East with a South wind laying our course for the Cape De Verde Islands which we hope to make in a week or two.

Wednesday Nov. 23rd —- Weather pleasant the wind having moderated toward morning, so that we could carry all sail. Have had another sea dish today, viz. potatoes, beef & onions in hash (by the way our beef for two days past & some time to come would make a landsman "heave up" it stinks so that it can be "heard" at masthead. Sail in sight this evening right astern.

Thursday 24th Nov —- Weather very pleasant, all sail set, sailing E SE with a SW wind. Spent the day as yesterday. Done nothing new save rigging a fore top striding sail. We have today had one of the most important of sailor's dishes. They generally get it twice or three times a week. I should judge that it was composed of flour, water & slush or pot-fat & boiled in a bag like pudding. It is served with very thin "lasses". It looks, when boiled, like nothing I can think of but itself, brown and solid. Hard a dish as it is, it is our best. Some think so much of it, that they will give a round price for another share. But, oh how I felt after it!!! For several days past we have had to be on deck all the time save six hours per night, 4 hours one night & 8 hours the next. This arrangement will probably last until we get the ship in trim & the work all done.

Friday Nov. 25th —- Weather same as yesterday, all sails set. Saw a flying fish for the first time, not near enough however to give a description. Occupied as yesterday. This morning the men were divided into boats crews & the boat steerers set about putting their craft in order. By craft they mean the irons used in capturing whales. Each boat has 4 harpoons, 4 lances, 1 spade, 1 boat hook & two or three pails for boiling. The irons are kept as keen as razors & each is provided with a wooden sheath to keep its edge & prevent accidents. It has been my luck again to get into the best boats crew.

Saturday Nov. 26th — Nothing new has transpired today. Everything is as yesterday save that the wind has shifted to the NE & a sail has crossed our bow but too distant to speak her. The crew begins to complain about our beef, it is so strong that one must hold his nose to get it down.

Sunday Nov. 27th — Weather calm. Hardly wind enough to keep steerage way. Spent the day reading & writing to Father.

Monday Nov. 28th — Weather pleasant with a good breeze which arose last night and relieved us of our fears as to the length of the calm in which we lay last night. While keeping watch in the early part of last night I saw the most splendid meteor I ever beheld. The old sailors say the wind always follows it in the same direction. Whether it always proves true I cannot say but it did last eve.

Two fin back whales were seen from the masthead. They are a species which on account of the danger in killing them are never taken. The greenhorns were on the *qui vive* until it was discovered that no chase was to be given. I think there will be some pale faces when we lower for the first time. I begin to find journalizing a difficult task. In a forecastle the shape of a triangle whose sides are 20 feet there are stowed 20 men, each has a chest and a pail of soap which nearly covers the floor. Everyone is obliged to eat, drink, dress, read, write & transact all his business upon his chest. While I have been writing here the Negro steward (a most important & impudent darkie) & two or three rascally greenhorns have kept our castle ringing with their lascivious songs & vile converse & one can well imagine what a place it is to write.

Tuesday November 29th — Weather still pleasant with a breeze that allows us to carry all the sail we have bent. Have stood a lookout for the first time today. I probably shall have to take my turn (trick we call it) hereafter. We were much amused today by a greenhorn who had to take his first lookout. After a tremendous effort he at length gained his position & there lay sprawled out very much like a half drowned pup on a bit of sail.

For two or three days past I have been thinking of stopping somewhere on the South American coast and go into some business if an opportunity should offer. I may alter my mind before I get there. If I do locate in that region when will my parents see me again?

Wednesday November 30th — Weather clear with a heavy breeze. Towards night it became so strong that we were obliged to take in top gallant sails & reef top sails. I find that there are greater cowards than myself. I have not as yet been in a position that gave me any fear. In the night we were ordered to furl the jib. Two besides myself laid out to do it. When we had nearly finished I spied one of the greenest sitting astride the bowsprit sticking to the cap for dear life. It forcibly reminded me of the manner in which a six year old would stick to the pommel of a saddle.

Thursday December 1st — Weather unpleasant. Raining frequently. Sailed under main & top sails until noon when the wind chopped around into the NW & fell off so that we could carry top gallants. Our course is & has been for some time SW.

Friday December 2nd — Weather squally. Wind high & sea rough. The spray dashes over the deck so frequently that we do not pretend to keep dry. They keep us green ones plaiting sennit. It is made from seven threads of rope yarn braided flat & used to wind ropes to prevent their chafing. While our friends in Yankee-land are blowing their fingers and wading through snow we are padding about deck barefooted with coat & hat off. A great consolation to us. The sailor seems to take delight, in tantalizing the fresh hands with visions of high living ashore. While sitting around our beef kid someone will ask for roast beef, biscuit or some other nice thing & if it don't make one think of mother's pantry then I've lost my memory since noon.

Saturday December 3rd –- Weather same as yesterday with a little less rain. Sailing S by E, sharp on the wind. "Larboard tacks aboard." By that expression we mean that the tack of the sail is hauled down on the larboard side. Nothing new has occurred today save that in placing a boat on its crane we stove a hole in the side. This was however soon repaired. The whale boat is a very tender affair when out of the water. It has to be handled like glass ware. When in the water there is no boat equal to it either for riding a sea or for speed.

Sunday December 4th –- Weather pleasant, wind fickle, which makes it necessary to take in & shake out sail frequently. I begin to believe that the sails are handled rather more than is absolutely necessary merely to "learn" us greenies how it is done. Spent the time today in reading & sleeping. The ship's small jobs are so near done that we have 12 hours below every day. Porpoises begin to swarm about the vessel. We shall probably soon have a dish of porpoise liver by way of change. We usually have beef and potatoes one meal & potatoes and beef the next. For variety beans, lobscosuse & duff…

Monday December 5th –- Weather pleasant. Wind fair & light. We have bent three new sails today, vis. main Royal, main top gallant striding sail & flying jib. As we are fairly in the trades we can depend for some time on fair & steady winds. Our course has been south with an Easterly wind. The Captain has been aloft looking for the Cape De

Verdes. We shall probably make them in a day or two. Saw a fore top sail schooner to leeward of us this morning sailing southwesterly.

Tuesday December 6th — Weather pleasant, wind full in the fore part of the day. This evening we lie entirely becalmed about 15 miles from one of the De Verdes. We came in sight of this Island early in the morning & have been lying near it all of the afternoon. The East, North & West sides of the island are barren rock to all appearances. Not a cabin in sight although the Captain says it is inhabited to the Southward. From the North the rock rises with a gradual slope to the right several hundred feet. It seems in some places to be cut by deep ravines & then it rises with lofty precipices. Not a specimen of the vegetable creation is to be seen.

This eve another of the cluster hove in sight. Brava is our stopping place here. We shall probably make it tomorrow. This forenoon spoke the ship *Leander* of Bristol, R. I. 22 Mos. out with 540 barrels bound for the Western Islands. Finished a letter today to Father & commenced another to Fish. The island we have seen is St. Anthony Lat.17° 12' Lon. 25° 19'.

Wednesday December 7th — Weather pleasant with a strong breeze. We left St. Antonio astern this morning & sailing S SE made the Fogo & Brava this after-noon. Fogo has the same appearance as Antonio save that it slopes more gradually to the Westward & looks as though, it might be inhabited. Brava lies to the SW of Fogo & is more level & even. Boobies & flying fish are about us in abundance. The Booby or Gony as it is called is a large sea fowl back black & belly white the bill is also white. It is frequently caught upon the spars having apparently no fear of man. When on deck it is impossible for it to rise. They are sometimes so large that their wings will reach from one bulwark to another. As we do not come to anchor here we are obliged to lie off &

on until the Captain has done his business. Tonight we shall have to tack ship several times. We didn't this afternoon for the first time while under full sail.

Thursday December 8th — Weather pleasant with the same old breeze. Last night we stood away from Brava until midnight & then ran back until 11 o'clock AM when we heaved to & sent a boat ashore with the Captain, 2nd & 3'd mates. The boat returned in about two hours with 3 black Portuguese as many of the crew having stayed on shore. The boat returned soon with 5 barrels of flour, some potatoes, tobacco & calico of the Captain's. The governor or collector of the Island came off & purchased 2 barrels of flour & some rice.

In person, the officer was about as much of a man in appearances as the poorest of our blacks in their holiday suits. His crew were dressed in shirts & pants. This is the dress of all the men on the Island. He that can raise a coat & hat is a rich man. When the boat came off at night it was loaded with melons, oranges, potatoes & pumpkins. Save a little that the men purchased the fruit all goes to the Captain. The men that went on shore sold everything they had about them for fruit or fowls. During the day we stood off & on, tacking the ship every hour almost. As soon as the boat was fast we shook out all our sails & stood on our course again. This eve we lie entirely becalmed a little to the leeward of the Islands. Almost every one of the crew left letters with the American consul here to be sent home by the first opportunity. We learned that the *Magnolia* left this but two days since. She sailed from Bedford ten days before us bound to the same cruising grounds.

Friday December 9th — Weather pleasant. All sail set. Sailing S by E. But little doing on board today save stringing up the pumpkins & braiding sennit. In consequence of eating four oranges last night I have been sick with the headache this PM. This eve the foremast hands had a chicken stew of the fowls they bought yesterday. Black fish raised today. Whalemen usually take them, why we passed them I can't tell.

Saturday December 10th — Weather & wind fair. Sailing South. This morn found a flying fish on deck. Was about six inches long of the shape of a trout & color of a shad. Its wings are similar to those of a bat but much more transparent. As we approach the line it grows warm very fast & it is now very uncomfortable below decks. The greenest of the green begin to fear a visit from Neptune. I rather think he will find customers in some of us.

Sunday December 11th — Weather & wind still fair, it continues to grow warmer. Course S by E. This Sabbath has been spent like its predecessors in reading & writing. I have continued my letter to Fish. A hermaphrodite brig in sight & making for us this evening. While I write the hands are requesting me to note the fact that 3 of the forecastle hands do the stewards washing for the cabin leavings.

Monday December 12th — Weather, wind & course same as yesterday. This morn one of the boat steerers tried his skill upon a porpoise but missed every time. We have worked harder today than upon any day previous. Some of us have been sheathing the deck & the rest breaking out the fore & main hold (or in other words, have been hoisting out everything inside the holds).

Tuesday December 13th — Last night the weather was squally with a little rain. The wind today has been minus, with the exception of an occasional puff, we have rolled at the mercy of the waves. Finished sheathing deck today. Sharks begin to swarm around us, they are not of the dangerous kind however. While writing here I have perspired more than ever I did in a week before. I find that my system is undergoing an entire change & in some respects I expect to boast a comfortable "corporation" when I return. 2 sails in sight today.

Wednesday December 14th –- Last eve we had a hard night of it. The first watch saw two or three poor sleepy chaps hoisted aloft by the heels and some others lashed together and to the belaying pins. The weather was calm and the moon shone bright & in such a time the bow of a ship is a jolly place.

The two last watches were kept in a tropical storm. The rain poured down in torrents & to better the matter we were obliged to handle sails & rigging most of the time. Through the day the weather has been pleasant & calm hardly keeping steerage way.

The mate caught a loggerhead turtle by throwing a harpoon through him from deck. From his flesh we had a glorious soup this eve by far the richest dish we have had in a month. Several of us tried a bath in the blue sea this eve among sharks & what we fear worse, an insect called Portuguese men of war, their bite is somewhat venomous though not dangerous.

Thursday December 15th –- Weather variable, sometimes perfectly calm then with a light breeze & occasionally a little rain. We have erected a blacksmiths forge today in a manner that was new to me. We took half a large cask & nearly filled it full of sand. On the sand with brick & mortar we made something that answers the purpose of a forge. Raised a whale this eve too late to lower for him. One of the hands lost his watch below & was obliged to pound the rust from one of the anchors in consequence of swearing in the Captain's hearing.

Friday December 16th –- Weather pleasant with a steady breeze steering nearly west. Spoke to an English East Indiaman the *City of Poulah* of London. She was a noble craft and as she swept by us with all sail set & everything taut & trim it made some of us sick of whaling.

Saturday December 17th –- Weather pleasant with a strong South wind. As South is our course, we have to tack ship occasionally in order to hold our own. Turned in this afternoon with the colic in consequence of eating part of a raw turnip.

Sunday December 18th –- Weather pleasant with an occasional shower. Steering E by S. Raised five sail today all at a distance. For a fortnight past our carpenter has been sick with the rheumatism continually growing worse. For 3 days our cook has been laid up. We do what we can for them but for being sick a ships forecastle is the hardest place in the world. A pile of leaves in an oak grove is a downy bed compared with it.

Monday December 19th –- Here I sit on my chest coatless & shoeless, every rag upon me thoroughly saturated with salt water & fresh. It has rained torrents nearly all the day & it is likely to rain all night. Tough as the weather is & muddy as is the forecastle floor, we are a right jolly set singing all sorts of songs & making all sorts of fun. We have at length a fair wind sailing S by W. Sail tonight ahead gaining on her fast, we think her to be the *Magnolia*. Captain says we are about crossing the line but it is not in sight yet.

Tuesday December 20th –- Weather pleasant since three o'clock last night. Last night we heard black fish blowing all round but have seen nothing today save two sail early in the morning. We have for the first time thoroughly scoured the floor of our pen. Before cleaning it resembled a hog pen in smell and appearance. We are now in latitude 4° North sailing E by N, losing ground all of the time.

Wednesday December 21st –- Weather squally until noon, pleasant this PM. I have for the first time in my life turned a spun yarn wheel. I believe that is purposely calculated for using up strength. If a Yankee dog could not invent a better I would disown him. It consists merely of a spindle & balance wheel run thorough the upright pieces of board made fast to the windlass bitt. It is turned by passing a short rope once around the spindle and holding one end in each hand. Steering E on one tack SW on the other with the wind South.

Thursday December 22nd –– Weather pleasant & calm. Hardly any steerage way. All hands on deck from sunrise to sunset setting up rigging. Evenings we lately have had a fair wind so that there is a prospect of our crossing someday. Saw a water spout at a distance.

Friday December 23rd –– Everything same as yester-day. Just came out of the water. It was much pleasanter than before. A school of Albacores & skipjack having destroyed the men of war.

Saturday December 24th - Weather pleasant in the fore & storm in the after part of the day. Making slow progress towards the line. Have overhauled our potatoes today. Have enough to last us this 3 mos.

While standing at the wheel last night I saw a curious phenomenon. From some cause unknown, the sea appeared to be but a bed of embers & the ripple at the bows a sheet of flame. The old ones say that they never saw so much phosphorescence at a time.

Sundry December 25th –– Last night was a fine one if thunder & lightning could make it fine. We were obliged to pull & haul all the time either taking in or loosening sail. Weather pleasant today but a dark prospect for tonight. A large school of porpoises crossed our bow this PM.

The Captain today offered 20 cents per barrel to him that raises whales. Have had a slight breeze & favorable most of the day --- sometimes calm. As today is Christmas we have a glorious duff by way of treat. The contrast between my situation this eve & a year earlier at this hour, however great, makes me feel none the worse for it. I rather prefer my present situation.

Monday December 26th — As I prophesied, last eve was a beautiful one -- it rained torrents from sunrise to sunset. We were obliged to handle sail as the eve before. Today it has rained most of the time. I have not been dry clothed this three days & probably shall not until we are well south of the line. Sailing S by W with a fair wind this eve.

Tuesday December 27th — Weather & wind fair, sailing SW by W all sail set. We are in hopes that we have reached the South East trades. Raised a whale this noon but were able to get but one sight of him. In wearing to sail for the fish I had the satisfaction of seeing my poor hat taking a lesson in navigation by itself.

Turned in a short time this PM with a headache but it's gone now. It is customary among all whalers to give the hands their regular watch below. But with few exceptions we have been obliged to spend the whole of the day on deck sometimes for almost nothing. If it continues thus much longer we shall leave the most of our crew at the first port we make. Set the second Main Royal today, the old one having gone to rags.

Wednesday December 28th — Weather pleasant & wind fair. Sailing SW with a whole sail breeze. Assisted in rattling down rigging. I think I can do it now as well as any of them. Whenever there is anything new to be done I generally manage to have a hand in it so that I shall the sooner be a seaman. Steward struck an albacore about 4 ft. long but lost him.

Thursday December 29th — Weather & course as yesterday. Wind a little fresher. Last night it blew a strong breeze, so strong that Neptune declined visiting us as we crossed the line. At work today as yesterday.

Friday December 30th –- Weather, wind, course, work & other things as yesterday.

Saturday December 31st –- Ships affairs same as yesterday. At work every day in the rigging. We shall soon have things ship shape aloft.

Sunday January 1st 1843 –- For a new year's day we have had a pleasant one, the first that I ever spent in bare feet, duck pants & shirt sleeves. We are now some ten degrees south of the line & fairly in the trades. Sailing S by W at the rate of 8 or 10 knots per hour with the wind SE. The Captain thinks we shall make the Horn in 20 days.

All sorts of skylarking this eve, some of the greenest have to take an overhauling occasionally. We have one right raw chap from Conn. with whom all hands seem to delight in fooling. Today is his birthday & he has an extra share.

Monday January 2nd –- Weather pleasant barring a sprinkle of rain. Wind comes in gusts, sometimes pretty strong. Have taken & bent a fresh main sail in order to prepare the other for the Cape. We have also taken out some new top sails for the same purpose.

Tuesday January 3rd –- Weather fine. Wind fair full & steady. Course S by W with SE. wind. Work today as yesterday, shifted main sails again.

Wednesday January 4th –- Weather fair save an occasional squall. Employed as yesterday. This evening while standing at mast head I had the satisfaction of witnessing a most splendid rainbow. It consisted of two arcs one in the sky & the other apparently upon the sea. The ends of both meeting about 30 rods from the vessel. All the different colors being clear & distinct.

Thursday January 5th — Weather pleasant. Wind this eve almost gone. This morn we came in sight of the isle of Trinidad situated in Lat. 20° 28' S & Lon. 29° 5' W. It is an uninhabited barren rock & appears at a distance much like St. Antonio. At noon we were abreast & now are far away from it.

While looking for the land this morn, the Captain fancied he saw a school of whales. After much hazing, hauling up sails, preparing boats & everything else that belongs to whaling, it proved that the Captain was mistaken. All hands hope that we do not see a whale this side of the horn, some say never. Course S SW with the wind in the opposite quarter. Could say today for the first time that I was under the sun. Tomorrow I shall be able to say for the first time that I have seen the Sun north of me.

Friday January 6th — Weather exceedingly pleasant. Wind light & variable. Course the same as yesterday.

Saturday January 7th — Weather hot & sultry, hardly a breath of air stirring save when a squall strikes us. Nothing now worth relating. Yes but there is! I have seen a large school of flying fish pursued by a large flock of sea fowl.

Sunday January 8th — Weather clear & pleasant but for the heat which fairly fries the tar from the deck. Light breeze this morn none this PM. Notwithstanding the days are so hot the nights are very cool.

For a month past I have made my bed of soft pine finding it more comfortable a berth. Last night I lay in the fore top to get out of hearing of all sorts of vile speeches. By way of novelty we had two regular fights in the forecastle, both bloodless however, as an instance of the love which a sailor has for the bottle one of the hands gave 2 dollars for as many quarts of liquor.

Monday January 9th — Weather hot & pleasant, just wind enough to keep steerage way. We have had an interesting time of it here this day. This morn the fellow that bought the rum on account of grumbling at washing decks was obliged to scrape the topmast.

When he came below he swore that he would have a regular drunk & in less than an hour was flat on his back kicking & striking at some imagined enemy. We picked him up & threw him into his bunk. As soon as there he made mincemeat of the berth above him. Drawing him from thence we bound him hand & foot, threw him into another bed & lashed him fore & aft. There he lay cursing the officers of the man of war he was in last. Myself he took to be the quartermaster be-cause of my binding him. At length he commenced cursing this vessel & its officers.

The Captain, coming forward, heard and called him upon deck. Instead of doing so he dared him into the forecastle & in he came, the mate at his heels. They seized & carried him upon deck, not however, until he had spoiled the Captain's shirt. They took him aft and lashed him there. Then coming forward he proceeded to search the chests but finding a bottle in the fellow's chest he gave it up satisfied. I believe that he has been let off with only the loss of three watches below. They must have thought him crazy. Saw some grampuses & porpoises this noon. Bent a new main top sail this P M.

Tuesday January 10th — Weather mixed, pleasant & squally by turns, sometimes quite calm. This PM from 4 o'clock to 11 o'clock was a hard time with us. Upon turning out at four I went to the mast head & there saw a brig (which proved to be the *Janas* of New York) bearing down directly upon us. Thinking it a good opportunity to send home I ran down & wrote a short letter to Father. I gave it to one of the hands & returned to my place aloft. I had hardly reached the cross trees when the maintop lookout sang out "there she blows" about a mile away.

Immediately the Captain ordered us to stand by our boats. This made it necessary for me to slide down the back stay in double quick time & make for my boat (the waist boat). The larboard boat lowered just before us. This boat has the best crew in the ship & everyone expected her to take the lead. Whereas we had the lightest crew (all save one green hands) & by far the worst boat in the ship. Although behind in starting, in less than fifteen minutes we were far ahead. As it happened a squall hid the whales & thus we lost them. As soon as we were fairly aboard again a very heavy squall struck us & made work enough but as the wind is fair we like it well enough. The row caused a sensation similar to sea-sickness in me.

Wednesday January 11th –- Weather pleasant. Wind fair & full. Rigged a lower studding sail & at this moment every rag is set.

Thursday January 12th –- Weather pleasant, wind light. Sea pretty high. Larboard tacks aboard sharp on the wind. Course S by SW.

Friday January 13th –- Weather clear & hot. Dead calm. Raised porpoises this eve. Broke out & trimmed the hold this PM. She has lately had a heel to port to accommodate her for the trades. Closed a letter to M. H. Fish.

Saturday January 14th –- Dead calm. Still a little rain this PM. Yesterday & today I have a touch of the rheumatism. Nearly gone. I hear that we are in Lat. 29° S. I can only 'say' from hearsay because we can only know from the cabin boy.

Sunday January 15th –- Weather pleasant with a good four knot breeze. Commenced a letter to father. Spent the time in reading & writing as usual.

The Log Book

Monday January 16th –- Weather squally. Wind full & strong, going at the rate of 8 knots per hour. As we are nearly off the river La Plata we are in daily expectation of tough weather this being the place for such things. This PM a large Liverpool Barque ran across our bows without speaking.

Tuesday January 17th –- Weather cloudy & cool, at least it feels cold to us who have left the tropics but a few days. We are now in the latitude corresponding to that of Connecticut & should have July weather but it feels more like October at home. Sailing our course with a good 8 knot breeze on our Larboard quarter.

Wednesday January 18th –- This AM weather pleasant & fair with a strong breeze. At noon just as the starboard watch were turning in for a snooze, a whale was raised close by us. All the boats lowered, ours taking the lead as before. We pulled all the PM, some of us after one whale & some after another as there proved to be a large school about us. At 5 o'clock we came on board with nothing to pay us for our trouble. We are now almost becalmed. We have about us several kind of birds. The Black Haglet, the Ring Eyed Haglet and the Albatross.

Thursday January 19th –- Weather pleasant. Calm in the morning, blowing at noon and calm again this eve. All hands are complaining about short rations. Yesterday I rowed five hours upon nothing but bread and molasses for dinner & one fellow had nothing.

Friday January 20th –- Weather clear with a strong breeze & heavy sea this AM. Milder this PM. Raised two fin back whales this morn thinking them to be right whales. Sent down the mizzen top gallant sail yard & mast & main Royal yard to be ready for squalls. Likewise we bent the fly jib & shipped in the boom. Had a fine job this forenoon scouring out try pots. When we come to such work as this we think we are about used up. Bent a new main top gallant sail & tried to strike a porpoise.

Saturday January 21st –- When I turned in last eve at 12 o'clock every sail was set & we were going ahead at the rate of 4 knots per hour. When I turned out at 4 o'clock we were under double reefed the top sails, fore & main sails. Before noon we furled the main sail & mizzen top sail & close reefed the fore & main top sails. Took in the waist and bow boats & stored them upon deck. Just before noon it blew so that the lee rail was some of the time under water. Suddenly it cleared away & now is fair & pleasant.

We are very fortunate in having fair winds most of the time so that we are seldom driven from our course. Weather grows cold fast but when I think of Connecticut as it is I feel warm enough. Tried again in vain to catch a porpoise this PM.

Sunday January 22nd –- Commenced blowing early this morn & continued to increase until at noon we lay too under close reefed top sail & fore sail in which situation we now are. The larboard anchor started & came near going overboard. Caught a dozen of Albatrosses with hook & line. Part of them we dressed & the rest let go again. One wing that I saved measured 4 ft.

Monday January 23rd –- Last night we carried all sail; this morn commenced taking in sail until we had reduced her to close reefed main & fore topsail & fore sail. This eve we are shaking them out again. The sea runs higher than it has at any time before but let the seas run & the winds blow I am ready to go aloft with the first. We prefer a gale to a calm at any time for then we have nothing to do but handle sail.

Tuesday January 24th –- Commenced taking in sail last eve at 3 o'clock & continued until 4 AM when the wind died away so much that we could carry all sail. We are now nearly becalmed. For the first time I had the satisfaction of raising a whale about five miles away. But Alas! It proved to be nothing but a fin back. Another blow was seen about five this P.M. The larboard boat lowered but made nothing by it. Porpoises & Genies in schools & flocks about us.

Wednesday January 25th — From a dead calm last night the wind arose by degrees & has increased until we have commenced reefing topsails. The day has been rather foggy & now it rains. Raised fin back three times, two of which I take the credit of. We are now in Lat. 42° 35' Lon. 47° 3' sailing our course at the rate of from 8 to 12 knots per hour.

Thursday January 26th-- From close reefed topsails of last night the wind has fallen off until it is now calm. Lowered the two boats on the crane this morning for a right whale but gallied him by crossing his wake. It is a curious fact that if you cross the wake of a right whale, he will see you and be off. Raised more finbacks. The Captain is now catching goonies having already a dozen on deck. As we run south the water is perceptibly colder, bringing woolen of all sorts into requisition.

Friday January 27th — When the Captain left the deck last night he left orders to set all the studding sails & Royal if a breeze sprang up. The breeze soon came & with it a hard two hours' work every rope was rove wrong & everything foul about the studding sails. When the larboard watch came on deck we had done nothing but get up the Royal & set it.

By this time the wind had freshened so much that instead of making sail it was necessary to take it in until nothing was left but main top sail & fore sail. One that never was out of sight of land cannot imagine what a storm at sea is. But when he has to go aloft with the wind howling all sorts of mournful music through the rigging or to stand upon deck when it lies at an angle of 45 degrees, the lee sail under & the water pouring in torrents over the weather thus he gets a taste of a storm.

The sea has been so heavy that it was impossible to walk from one end of the ship to the other without holding by something. A little poetical fiction would call it mountains high but I should call the swells mere knolls. When the wind was blowing its hardest this morn the weather Martingale guy parted but was soon repaired. Have seen several whales from deck there being no lookout kept in weather that is too rough to lower the boats.

Saturday January 28th –- The first watch of last night was very stormy, not one of us could boast a dry jacket save the "sojers". By noon we could carry top gallant sails. At 4 o'clock the top sails were double reefed. At 7 o'clock all hands were called to take in sail.

With much grumbling we went at it & by the time we had the main sail furled, a regular Cape Horn squall struck us & amid the worst of it we had to furl the jib, fore & mizzen top sail. However it is all done now & no one hurt & but a few frightened. Raised several fin backs. At 8 o'clock I must go on deck. The squall is over & there is room for another.

Sunday January 29th –- Weather same old sort. A squall every hour almost & when we are not in a squall we are laying to under close reefed top sail & foresail. All hands are occasionally upon deck to take in sail if any happens to be loosed. The wind howls mournfully among the ropes & occasionally we get a little hail.

Monday January 30th –- Lay to last night under close reefed main top sail & fore sail. This morn commenced making sail & by noon had everything out. In shaking a reef cut of the fore top sail a point was left untied and the consequence was a rent sail. We soon bent a new one. Have moved all the spars & lashed them amidships. Raised a fin back. Sail in sight. Suppose her to be the *Coral* of Bedford on the same errand with us.

Tuesday January 31st — Calm all last night. A fair gentle breeze through the day. The sail we saw yesterday lay nearer to us this morn & about ten o'clock the Captain of the *Coral* & a boats crew rowed 20 miles to overtake us. Soon after he came on board our mate and a boats crew went to "gam" with the *Coral*. Towards night we hove to, to let them overtake us, they being unable to do it without. At sundown the Captain returned promising to bear us company.

Wednesday February 1st — Weather fair in the AM & this PM a heavy mist is falling. The *Coral* that we expected to lead us was far astern this morn & now is further still. The men that went on board her yesterday tell such hard stories of the officers & crew that we are contented to remain in the *Nimrod*. They told of one fellow that was called down from aloft & made to eat a piece of meat that was taken from the dirt bucket. Some of the officers are frequently trying their skill upon the porpoises that swarm about us but as yet have had no success.

Thursday February 2nd — Weather last night & this AM exceedingly foggy & wet. This PM more pleasant. Since Tuesday morn we have sailed our course SW by W with the wind aft & on the Starboard quarter. This noon the wind shifted to the SW. We are now running W by S under double reefed topsails expecting a storm. The *Coral* is now six miles astern. I believe we are between the Falkland Islands & the main else soon will be.

Friday February 3rd –- Weather clear & cold with a whole sail breeze. Sailing S by E on the starboard tack 13 points away on the other. The water grows green an evidence of the vicinity of land.

Saturday February 4 –- Weather clear, cold & blowing. Running under double reefed top sails till night. Now with all sail set. We are beating against a head wind hardly holding our own.

Sunday February 5th –- From two o'clock until 10 this morn we were sailing before the wind with studding sails set. Suddenly the wind hauled from North to West & from that time to this it has blown a "screamer". When the first squall struck us we had to call all hands to take in sail for the first time since we have been out. This AM there was a mist so heavy as to wet us through & this PM every squall brings a shower. The Captain thinks that the Falkland Islands lie to leeward of us & in sight. But for this we should have comparatively easy times, as it is we have to clew up & sheet home the top sails every fifteen minutes as the wind blows. A sail between us & the land endeavoring to run clear of the land by carrying a press of sail. How she will weather it is uncertain as yet.

Monday February 6th –- Last night was the most uncomfortable that I ever saw. The wind continued to increase until we took in everything but close reefed main top & fore sails. Since daylight the wind has fallen off until we can carry main top gallant sail.

While standing the anchor watch last night the ship took a sea that made everything start again & what was worse, I took the brunt of it. If I am any judge of quantity at least three hogsheads fell upon my head & back. Drenched as I was, I was sent upon the fore yard to look out from that place. When I was relieved I could hardly move. The wind is nearly S SW & we are running upon both tacks.

Tuesday February 7th — Last night the wind shifted around to the Northward & fell off until at 8 AM it was dead calm. At 10 a light wind sprang up from the south but now it is entirely calm again. While at mast head today I saw a large school of sperm whales about five miles away but held my peace it being our determination not to raise anything this side of the Horn.

There are more birds of different kinds floating & flying about us today than I have seen at any time before. I have seen today Mother Carey's Chickens, three or four species of the Right Whale bird, the Albatross, Haglet & a new bird called the Mollymock. It is nearly the shape & size of the Goney but smaller. The wings and the back between the wings is black the rest perfectly white. It is amusing to watch a flock of chickens in a calm day. While feeding, they never rest upon the water but hop about upon its surface very much as a chirping bird would upon the ground. They are less in size than a newly hatched chicken, perfectly black save a small spot of white near the tail.

Wednesday February 8th — At 10 last night a gentle North wind sprang up & has held on until this time wafting us on our course at the rate of 4 knots per hour. We are now in Lat. 53° 25'. If this breeze holds us three days more we shall run clear of the Horn. Weather very pleasant for this latitude. The multitude of birds that were about us last night are every one of them gone.

Thursday February 9th —The fair wind of yesterday lasted us until 10 this AM when it shifted to the southward. It is now hauling back to the West again. This morn we came in sight of Staten Land & are now abreast of it. With a fair wind we shall now commence doubling the much dreaded Cape. Sail in sight sailing in the opposite direction to us.

Friday February 10th — Wind fair & fresh from the North sailing South by East. This morn raised Staten Land, that which we saw yesterday being something else. We are now so far south that we shall soon turn to the westward if this wind holds.

Saturday February 11th — Weather misty & rainy at times. Wind very light, sometimes fair but now it comes from the SW & we are running SE. The swell we have here is the longest & heaviest that I have seen. Eight of them will measure a good mile. Have seen a curious concern, half fish & half fowl called Waugen. The cry it utters much resembles that of a person in distress. This is doubtless the beast that gave rise to the existence of Mermaids in this region.

Sunday February 12th — Clear calm & comfortable. Nothing to do but read & brace yards. Sail just in sight to the northward.

Monday February 13th — Fresh fair breeze from the North & East held us in our course (W by S) until 4 o'clock PM when it died away. Another has now sprung up from the East. For several days we have hardly held our own having but wind enough to stem the current which runs at the rate of 30 miles per diem round the Horn.

Passed three sail, one of which, the *George* of Bedford, spoke & came on board of us. She is 38 months out with 1500 barrels of sperm oil. Lost her Captain, mate & four hands of the scurvy. Lost every sail twice in typhoons besides a boat & some of her spars. She is commanded by a Captain that lost his vessel at Owyhee some time since. Sail gone out of sight ahead of us this PM. Saw a fin back aloft this PM. Started a letter to Friend Fish in Salisbury by the *George*.

Tuesday February 14th — Last night was foggy & almost calm. This morn the fog cleared away & left rather a black looking sky. By the time that we had made everything snug it commenced blowing from the SW, but not so hard but that we could carry whole top sails until 5 P.M. when we double reefed the top sails & clewed up the mainsail for a squall which has not as yet appeared but on the contrary we have shaken out some of the reefs & loosed the M.T.G. sail. Wind fairer tonight. Snow & hail enough fallen to have a game at snow ball. Spoken the ship *Mary Ann* of Sag Harbour bound upon the North west coast with us Lon 63° 42' W.

Wednesday February 15th –- Weather clear, wind fresh & fair from the North & East. Sailing W by S with every rag out. Have to keep a look out forward for icebergs day & night but have seen none as yet. Unwell but not off duty with the headache. Sailing in company with the *Mary Ann* neither ahead as yet. Lat. 58° 12' Lon. 67° 20'.

Thursday February 16th –- Until 4 PM. Weather clear with a strong breeze from the E & S. Every sail set. Studding sails out below & aloft. At 4 it freshened so much that we had to double reef top sails & thus are running rail under. Steering W but our actual course is NW by W owing to the variation of the compass. We have now doubled the Horn with studding sails out, a thing that is hardly ever done. The *Mary Ann* is out of sight astern tonight.

Friday February 17th –- Running under close reffed topsails & fore sail, steering W NW sharp on a south-wester. Seas higher than ever before I have seen them but not 20 ft. as yet.

Saturday February 18th –- Wind commenced falling off last night & before breakfast we had everything out, even top Gallant studding sails & the wind aft. The wind has since hauled round to the N & W so that we are now running SW with a fresh breeze & thick fog. Sail in sight 6 miles to leeward to night think her the *Mary Ann*.

Sunday February 19th –- Blown a gale since last night from the N. Running west under closed reefed top sails & foresail. Just rain enough to keep us wet & uncomfortable.

Monday February 20th —- Wind fell off last night so that we have carried T. G. sails. Tacked ship at 2 PM & are now running on the wind N NE. Until today for some time past the two after boats have been turned up to prevent their getting stove by the sea in rough weather.

Tuesday February 21st —- Last wind freshened until we reefed the fore & mizzen top sails after which it fell off until we could carry top gallants. Tacked twice & the last time headed N which course we have held. Weather cold & cloudy. As we run N the days grow shorter. Off the Cape we had 18 hours of daylight, night lasting only from 9 PM till 3 AM. The sun instead of rising to the northward by compass, rose in the SW & set in the SE. Raised a fin back from deck.

Wednesday February 22nd —- Westerly wind strong & fair. Sailing N NW with all sail till 4, since under topsails & courses. Ran through several squalls without starting tack or sheet. Saw a fin back whale from aloft.

Thursday February 23rd —- Weather squally, cold & uncomfortable. Wind W SW running NW by N under double reefed top sails & courses. Lee rail under at that some of the time. Saw a very large fin back only a ships length off from deck.

Friday February 24th —- Last night weather squally & rough. Since morn the wind has fallen off & astern until we are just waging along before it. Spoke the bark *Adolph* of Hamburg from Valparaiso bound to Buenos Ayres.

Smith, the fellow that had a drunk some time since got into another scrape today. Under a pretense of sickness he lay in all last night (a thing he is in the habit of doing) & when the Captain called him on deck he hit him a slight tap on the head as he came out of the gangway for which he received a blow in the face & from thence sprang a row that has given the fellow a pair of irons, rations of bread & water & a berth in the sail room. Fin backs again.

Saturday February 25th –- Strong S wind running SW with studding sails below & aloft at the rate of 10 knots per hour. Commenced clearing the deck by replacing the spars that we had secured amidships for safety. Two sail crossed our stern this AM bound round the Horn.

Sunday February 26th –- Wind lighter than yesterday otherwise the same. Raised a sperm whale. Lowered three boats but missed him. The Captain gave the mate a sound scolding for not knowing his signals.

Monday February 27th –- Wind S, course N by W. Occasional showers, weather comfortable. Rigged out the fly jib boom.

Tuesday February 28th –- Weather, wind & course, same as yesterday. Saw a spout this morn but made nothing of it. Spoke the ship *Navy* of Newburyport, 32 months out, 2000 barrels. Sent our carpenter home by her having been unable to do his duty but a few days since we have been out. Smith was relieved of his shackles this morn. Sent up the Mizzen top gallant mast yard & sail this PM.

Wednesday March 1st –- Weather same as yesterday. Wind on the other quarter. Raised the Islands of Massa Fuero & Juan Fernandez this morn, the latter we are now abreast of & fast passing. Raised a breach just at night.

Thursday March 2nd –- Weather & wind same as yesterday. Course NW. Have been making preparations for taking in oil if we get any.

Friday March 3rd –- Weather, wind & course same as yesterday. We are so far north (Lat. 28° S) that Cape Horn clothing is very uncomfortable.

Sail in sight. Placed the bow boat, newly painted, upon her cranes for the first time this side of the Horn.

Saturday March 4th —- Weather wind & course same as yesterday. Nothing new save breaking out & preparing the holds for taking oil.

Sunday March 5th —- Course same as yesterday. Milder weather, hotter. Raised a breach but nothing but a breach. Lat. 25° S.

Monday March 6th —- Weather cloudy. Wind variable, mostly SE. I believe we are in the trades. Caught a porpoise last night it being the first blubber that has been brought aboard. The porpoise is a singular thing neither fish, flesh nor fowl. It has the body of a fish the flesh of a bullock the liver & entrails of a swine every part was saved. The brains went to the boat steerer that struck him. The blubber was half an inch thick & taken off in strips. Lowered for black fish but caught none.

Tuesday March 7th —- Weather cloudy with a little rain. Wind variable & changeable.

Wednesday March 8th —- Weather pleasant & warn. Wind light & sea smooth. The Captain says that this is the kind of weather that they always have in this region. He says he has been here three months without a blow or a drop of rain. Lat. 20°

Thursday March 9th —- Weather, wind & course same as before save a little less wind sometimes. Was roused out before breakfast to lower for black fish. After a short pull the waist boat (mine) fastened to a large one. As soon as the iron struck him he sounded taking out 30 or 40 fathom of line. Upon rising we gave another iron & lanced him a few times when he turned upon his side, fin out. We towed him aboard, hoisted him on deck and tore off the blubber, some of it 4 inches thick.

Minced & now are trying it out. We expect to get three barrels. In getting into the boat I sprained my foot so badly as to render it useless.

THE ATTACK

Friday March 10th — Weather, wind & course same as before. Tried our blubber out last night & made 4 barrels of it. A good quantity for a blackfish. Off duty by my foot. For several nights past a comet with a very long, luminous train has been "chasing the sun" until it has very nearly overtaken it.

Saturday March 11th — Affairs as yesterday. Nothing new.

** **Sunday March 12th** — Weather etc. continues the same. Lowered this noon for black fish. The larboard boat brought a small one aboard. Raised a whale spout this eve. Unable as yet to do much with my foot.

Monday March 12th — Weather etc. the same. Raised a blow which proved to be a fin back. Thus far I have raised more whales than any foremast hand. Commenced scraping ship preparatory to painting. As we approach the Galapagos Islands we daily discover some new species of birds. At present we have about us the Man of War hawk, the Booby

& Marling spike. The first, a large black bird that commands the seas or the feathered race thereon. The second a white bird about the size of a barnyard fowl with very short wings & slow in movement. The last is about the size of a robin, white & possessed of a slim spiral tail about 10 inches long. The bill is usually red.

NOTE: Evidentially a day has been lost here as Sunday and Monday are both dated the 12th.

Tuesday March 13th –- Weather clear & calm. Lowered boats just at sundown for black fish. The waist & larboard boats brought each one aboard.

Wednesday March 14th –- Weather same as before with a slight breeze from the NE. Raised sperm whales this morn. Lowered at 9 o'clock & pulled until night without any success. We several times got almost near enough to strike but they were too quick for us.

Thursday March 15th –- Weather same as before save a rain squall this morn. Lowered for whales again at 10 o'clock. Struck but drew. Lowered again this PM & the larboard boat struck and got one. After they were fastened we went up and gave him another iron & then came the melee. All four of the boats rushed on & then such a jamming; such a howling. It was stern all & pull ahead. Lay this way & lay that. In the end however no one was hurt, although his flukes came so near that I could have grasped them. As soon as alongside we commenced cutting in but did not do much. Tonight we are lying too under reefed top sails standing boats crew watches.

Friday March 16th –- Weather pleasant but hot. Commenced cutting in again this morn & by 4 o'clock had all the blubber below & the try works in operation. Finbacks all round.

Saturday March 17th — Occasional squalls of rain. Finished trying out. Find she makes 40 barrels

Sunday March 18th — Clear & pleasant. Course NW by North. Wind aft. Lat. 4° S. Finbacks all round.

Monday March 19th — Clear & pleasant. Course NW, wind NE. Stowed oil between decks & scrubbed deck. Fin backs again. Saw a large log afloat near us. Took it at first to be the head of a boat.

Tuesday March 20th — Weather fair. Wind shifted around to the S. Early this morn raised the Galapagos Islands & have passed several of the group since. We are now running to the Westward fast, leaving all signs of land. We are again north of the line.

PULLING TEETH

Wednesday March 21st — Wind light. Weather clear. Running W near & nearly parallel with the line. Painting ship outside.

Thursday March 22nd — Weather same as yesterday save the clouds tonight. Course W SW. Since we have been on the line the wind has invariably has been fresh at night & nearly calm by day. Commenced tarring down all the rigging for the first time.

Friday March 23rd –– Weather, wind & course as before. Been tarring rigging again. Fin backs.

Saturday March 24th –– Weather hot with an occasional cooling shower. Course SW, Wind S SE. Sail in sight.

Sunday March 25th –– Weather etc. as yesterday baring the rain.

Monday March 26th –– Weather warm & pleasant. Course W by S. Wind strong S SE. Finbacks. The comet I mentioned some days since instead of losing itself in sun light like others, has had a retrograde movement apparently going tail foremost & losing itself in the distance it is now quite dim. Its tail ends in the south side of Orion & its head is some two degrees to the westward.

Tuesday March 27th –– Weather & course same, wind somewhat fresher. Raised whales close alongside, lowered & just at sundown the Captain struck a lazy one & by dark he turned up. By 10 PM we had him alongside sail taken in & men below.

Saturday March 31st –– Just as on Wednesday, we were preparing to "cut in" and the fluke chain parted & the whale went adrift. After much trouble & hard work we secured him again & by night had him cut in & the try works in operation. Thursday we continued trying with the rain pouring in torrents. Friday we had the last of eighty barrels in the pots just as the sun set & today we have been getting things & ourselves in trim again.

The Log Book

Sunday April 1st —- April 1st in the States but the 2nd in this region. Weather pleasant, wind fair. Course W by S. We lowered this morn for a lone whale but lost sight of him.

Monday April 2nd —- Weather & wind same. Course W NW dead before the wind. Lat. 4° 42 ' S – Lon. 118° W.

Tuesday April 3rd —- Weather, wind & course same. Been hard at work all day at stowing down the oil we have taken.

Wednesday April 4th —- Weather & wind same. Course NW, raised whales twice but made nothing of them. All hands on deck finishing stowing down.

Thursday April 5th —- Weather etc. the same. Nothing new.

Friday April 6th —- Hot & calm, lying close upon the line. Fin backs. A general row today about a spinning winch that has been thrown overboard by some rascal.

Saturday April 7th —- Calm & hot, bent an old fore sail & M top sail in the place of new ones that were up. Saw today a curious monster called the Diamond fish as it lay upon the water its form was triangular, it sides at least 3 feet. Upon one side there projected something similar to a pair of horns. Accompanying it were two milk white suckfish about two feet long. Fin backs.

Sunday April 8th —- Wind light, course NW. Nothing new.

Monday April 9th —- Same as yesterday. Employed in shifting & repairing sails taking down good & sending up poor.

Tuesday April 10th –- Weather cloudy with a little rain & wind little fresher. Course NW by W. Finished shifting sails.

Wednesday April 11th –- Weather unpleasant in consequence of frequent heavy showers. Wind occasionally pretty fresh & again almost calm. We have finished nearly all of our work upon the rigging and are now employed upon small jobs such as making nettles & frig lines, alias very small cord. Closed a letter to father & commenced one to M. H. Fish.

Thursday April 12th –- Occasional showers, we have passed through the Doldrums as the old cruisers call this region of variable winds & showers & are now in the southern edge of the NE Trades.

Friday April 13th –- Weather cloudy wind strong Making at least 8 knots per hour. Course W NW.

Saturday April 14th –- Wind NE & very strong. Took in all the light sails last night & set them again at noon save fly jib. Light showers occasionally. Course NW by W. Running at least 3° per day.

Sunday April 15th –- Weather cloudy, wind NE., course NW. Wind so strong that we can just carry Main Royal & T.G.S. Sail. Got up the main T. G. preventer brace. I believe the skipper means to see how fast he can run the old *Nimrod*.

Monday April 16th –- Weather, wind & course same as yesterday save that we can carry nothing above top sails.

Tuesday April 17th –- Weather unpleasant. Wind attended with some rain comes in strong gusts. Not as steady as before. The Captain thinks we shall come to anchor by Saturday. Blackfish all around.

The Log Book

Wednesday April 18th –- Weather pleasant. Wind just NE & mild enough to use T. G. S. Sail. Course N NW. Although within the tropics the weather is cold enough for October at home. Employed in making fancy gaskets and other small affairs to make the *Nimrod* appear well in port.

Thursday April 19th –- Weather, same course W by N. Employed in scraping & painting the bulwarks & spars & getting casks ready for water.

Friday April 20th –- Weather pleasant, wind lighter, course same. Employed as yesterday. Expect to see Maui tomorrow morning. Sail in sight. Closed a letter to Fish.

Saturday April 21st –- Weather warmer, wind lighter. Course W & for the NW point off Maui alongside of which we are running, waiting for daylight to come to anchor. We raised the land at daylight this morn & so lofty is it that we have just reached it running 4 knots per hour. Employed in getting up Royal yards & polishing up little things.

Sunday April 22nd –- Weather pleasant. Although close to land in the morn yet between calms & contrary winds it was 3 o'clock before we came to anchor. The anchoring ground is on open roadstead on the S side of the island & between Maui, Raim & Molichi. We found about 24 ships already at anchor & boats plying between them. A dozen have been aboard us now.

Went ashore a few minutes with the Captain & of all wild looking chaps, those kanakas take the lead. The huts are built of turf & mating covered with a kind of grass. Some however are handsomely built in English style. The inhabitants (or natives) are many of them dressed in English style principally in clothes purchased of sailors. But many yet dress in the native style which is merely a tapper or narrow piece of

39

cloth wound about the loins & a piece of native cloth thrown over the shoulders. Got track of Mr. Damond who is now at Maui on a visit. His station is at Oahu where he resides.

Wednesday May 3rd — We are now 2° north of the Sandwich Islands running north close hauled upon a strong NE wind bound straight to the NW.

We weighed anchor yesterday morn after having gone through the usual routine of marvelous events. The men had 3 days liberty apiece, one watch going ashore one day and the other the next. Each being obliged to come off at sundown or be put in the fort by the Kanakas, which fate five of the men suffered. The ugliest of the crew was left there in durance vile.

One of our men ran away & could not be found & the cook undertook to get away from the ship in the deck tub having all his clothes stowed in a smaller tub. He got his vessel & longboat well afloat but in getting into the larger tub it capsized & he was compelled to call for help. When he got on deck again all he had left of a good outfit was the shirt & pants he had on.

While ashore I saw many things new & curious, especially in their manner of living. They retain many of fair old customs yet & for convenience are like to. They raise large quantities of sweet & Irish potatoes & Taro, Bananas, Coconuts, breadfruit, sugar cane & various other kinds of tropical fruit is raised in abundance.

I saw the Hawaiian King & Queen, also my former preceptor Mr. Damon with whom I had a pleasant visit on Sunday. I heard him preach at the seaman's chapel.

Monday was the day appointed to sail but the wind was too strong to get under way with safety. Almost every vessel dragged her anchor; early in the morn a large Frenchman came into us carrying away his round house spanker from a gaff without doing us the least damage. In the afternoon another ship dragged into her carrying away her spril sail yard & her own fly jib boom.

We made ourselves secure by dropping both anchors & paying out about 130 fathoms of chain in 20 fathom water. Short as was our stay in this place it was long enough to sicken me of shore & fresh grub & it pleased me much to let go the bunt of a topsail again. Although we have potatoes & sometimes fresh meat yet, I take my salt horse & hard bread in preference to anything else. We shipped three natives to fill up our crew & they are not quite as bad as I was when first out.

Thursday May 4th –- Wind still fresh. Course full and making N. Weather quite cool.

Friday May 5th –- Having just got out of the NE trades the weather has moderated much, it being now quite comfortable. Course full & making N by E.

Saturday May 6th –- Weather pleasant, wind milder. Course same. Nothing new.

Sunday May 7th –- Weather & wind same. Course NE by N. Starboard tacks aboard. Wind abeam. At 2 o'clock PM lowered for a large sperm whale and did not get on board again till nearly 9 o'clock. For the first five hours we pulled to windward with all our power without a moments rest. Although we had a green Kanaka in the boat that was worse than nothing, we passed all the other boats and left them far astern.

Monday May 8th —- Weather pleasant. Course NE by E. Wind strong on the S quarter. The Albatross begins to make his appearance.

Tuesday May 9th —- Rain pouring in torrents nearly all day. This forenoon the wind came out ahead so that we are making W by N. This eve double reefed the topsails.

Wednesday May 10th —- Last night it continued raining very hard. Cleared away this morn. Since it has been very cold & rough. Wind same. Running NW by N & NE by E.

Thursday May 11th —- Been taking in sail until we are laying too under closed reefed M top sail, M Spencer foresail & F T Stay sail. Sea exceptionally rough. Wind from the NW. This is just the kind of weather that forecastle men wish for nothing to do but "sojer."

Friday May 12th —- Weather pleasant. Wind light. Course NW & NNE. Sharp parted & repaired mizzen stay.

Saturday May 13th —- Almost calm since last midnight. Heading nearly N. Caught a small sun fish weighing nearly 200. Lowered this PM for a spout but after pulling two miles could find nothing but Grampuses. We are now in a most uncomfortable latitude. The nights are so damp as if it was continually raining by an exceedingly dense fog which always exists at night in this region.

Sunday May 14th —- Weather pleasant. Course N by W. Wind just abaft the beam. Starboard tacks aboard. For some days past there has been a vast quantity of Barnacles & Portuguese men of war floating upon the surface of the water, every day growing thicker. Today nearly one tenth of the surface is covered with them.

The Log Book

Monday May 15th — Weather pleasant. Wind fair & light. Course N by E, S tacks aboard. Some of our men begin to feel the effects of their intercourse with the Sandwich Island natives.

Tuesday May 16th — Weather pleasant. Wind on S quarter. Course N by E. After 6 months hard work upon our rigging we have at length come to a stand. We have 20 fathoms of sennit to make to each man & then we are done for the present.

Wednesday May 17th — Wind, weather & course same. Nothing new.

Thursday May 18th — Weather foggy till noon. Since nearly clear. Wind strong on S quarter. Course N.

Friday May 19th — Weather foggy & since noon the wind has died away but still remaining S. Course N. Last night it blew almost a gale & being before it she rolled heavily. At 9 last eve she gave a roll that filled & carried away the outer bow of the waist boat with some of her trimmings. All hands were called to take in the bow boat, turn up the remnant of the waist & shorten sail. Today returned the bow, took in & commenced repairing the waist boat which is quite a job.

Saturday May 20th — Weather exceedingly foggy enough so to keep us wet "entirely". No wind, therefore no course. Now that it is too wet to work on deck, the officers to keep us from spoiling, have given us a job at picking Oakum. God bless them for the care they take of us.

Sunday May 21st — Foggy in the morn. Pleasant through the rest of the day. Wind light. Running N by E. Larboard tack.

Monday May 22nd — Weather cool but very pleasant and almost calm. Heading N close hauled on L tacks. Blackfish & Grampus in abundance. Took down the Fore top sail to repair at 10 o'clock.

Since writing the above which was done at Sundown, we have raised a right whale killed him & have him alongside with all the sail save M top sail taken in. The twilight lasts so long that it was nearly all done by daylight. All the boats were fast save the waist which is repairing & the officers say they never saw one killed so easy.

Monday May 29th — Since writing the above, a week has passed & we have in that time tried, cut & nearly stowed 300 barrels of Right whale oil. Just as we were finishing the first whale (in the doing of which we had a capital time). On Thursday we took another 150 barreler & now are running him down below.

Tuesday cut in & lowered once. Wednesday fair weather. Right whales all round. Lowered several times. Thursday fair. Lowered the larboard & bow boats after dinner. The bow fastened. Lowered the starboard boat (in which I was compelled to go as bowsman) & fastened but for a long time was unable to hit his life with the lance, our Captain being no right whaleman besides having no good opportunity. At last the mate, a man at home in this business, got one lance at him & set him to spouting blood by the barrel which soon turned him up.

Friday cut in his head & lips & commenced trying. Saturday obliged to stop the works on account of weather being too rough to boil. Commenced again at night & now are done, the weather being quite moderate.

Such a looking set of fellows as we are, few landsmen ever saw. Our faces black with smoke, our clothes all oil & gummy from head to foot. The most of us have sore hands for every little scratch becomes a bad sore when once it gets raw oil into it. As for myself, I have a fellon that uses up one hand & a raw sore upon the other.

The Log Book

Tuesday May 30th –- Weather pleasant in the AM. Misty in PM. Lowered this AM but could not get near the fish. Been running down oil. Nearly through. One side of the main hatches is full. Just at night saw two right whales within a quarter of a mile of the ship & there lay until dark. Commenced standing boats crews with all sail in save close reefed M top sail, fore & fore top mast stay sail & M Spencer. Blind Jack at the wheel.

While taking in sail one of the men fell from the bowsprit, but being a good swimmer, caught one of the sideboards & saved himself. For the last week we have remained nearly stationary in about Lat. 50° N & about 6° from the American coast not more than 3 days sail from Astoria.

Wednesday May 31st –- Weather cold. Wind N & very strong. Beating to windward on half day tacks. Seen but one spout. Been cleaning out between decks & preparing for another Fare.

Thursday June 1st –- Weather fair. Wind light. Lowered twice, the second time we ran afoul of 3 whales "in a game". The Captain fastened & then the other boats in turn. The moment they were struck they all "brought to" & commenced lashing their flukes about in all directions, working their unwieldy carcasses as nimbly as a skipjack.

Our boat was one of the first in & in five minutes Mr. Sanford killed his first whale with his second lance. Another we all tried in turn & together without avail. At length the frightened monster in company with his unscathed mate started to windward & the way they walked off five miles with two boats in tow would astonish a locomotive.

By spading the flukes of the whale to which we were fast we again brought him to & commenced upon him anew, but it was not until after he had received a dozen irons & thirty lances that we could hit his life. Mr. Smalley at last did for him. While the three were together one of them ran his head against our boat & gave admittance to a little salt water through the cracked seams.

But that a Right whale always changes his course as soon as his head meets an obstacle, our boats crew would have inevitably been "spilt in the drink". When the last whale turned up we were at least five miles from the ship & directly to windward with a light breeze. Therefore the whole distance must be towed by two boats crews, the other two vying with the other whale a mile or two to leeward of the vessel.

Every man of us was wet, without jackets & without food save a little biscuit, but the whale must go along side & at it we went, pulling all we could from seven PM until two next morning. Finally after fifteen hours of the hardest kind of work we had our fish fast to the ship & the other waited unable to get him alongside. While coming on board the third whale kept close to us rising every few minutes within a stone's throw apparently altogether unconcerned. We should have killed him but that two were enough at a time.

Friday June 2nd — Weather fair, wind light. After three hours sleep we were this morn called to get our other whale alongside & commence cutting in. Finished cutting at 8 o'clock & all hands were permitted to turn in for eight hours. Two sails in sight. Whales blowing all round.

Saturday June 3rd — Wind fresh & cool from the North. Cut in our other whale & finished about 7 o'clock. Boiling all day. Owing to some disease about the whales head we lost three fourths of the bone.

Sunday June 4th - Wind & weather same. Trying & preparing to stow down. While breaking out the after hold a large cask swung across the hatch & jammed my leg somewhat badly.

Monday June 5th — Weather & wind same. Boiling & stowing.

Tuesday June 6th — This morn weather fine. This eve blowing a gale & raining torrents. So rough that we have to cool down just before finishing. A very large whale came so close to us as to be gallied by her.

Wednesday June 7th — Weather same; fairer yet rough with some rain. Finished boiling & stowed down 90 barrels. Lying too under closed reefed M top sail. Blind Jack at the wheel.

Thursday June 8th — Wind SW & strong. Sea rough, too rough to lower sail. Sail five miles to leeward lying to like ourselves under close reefed M Top Sail & Fore Top Mast Staysail. Finished stowing down a fare of more than five hundred barrels of Right whale oil which, with our sperm makes us about 650 barrels Stowed the most of our bone among the oil.

Friday June 9th — Wind Northerly. Weather cold and damp. Lowered this morn but missed our whale. Soon after spoke the ship *America* of Stonington ten months out with 600 barrels. We lowered at noon. Larboard boat fastened & had her line run out in less than five minutes. Bow & waist boats fastened soon after & gave him 3 lances. He then sounded, never again to rise taking all the bow boat's line and half of ours. We had to cut to save the other half. Just at night the *America* ran down to us & sent a boat aboard us to get some medicine for two men that got badly hurt by a whale today.

Saturday June 10th — Wind E. Weather too thick to lower. Saw a spout this morn. After scrubbing decks all hands turned in for the rest of the day save when wearing ship at 4 o'clock.

Sunday June 11th –- Weather as yesterday until 4 o'clock when it cleared off & sail was made. In an hour we lowered and fastened & had to cut a fouled line. Was unable to fasten again.

Monday June 12th –- Wind E & strong. Cleared away just long enough to lower once but made nothing. All hands below most of the time.

Tuesday June 13th –- Weather cleared. Foggy at intervals. This morn the Captain turned the ships head S again (having been running N for three days past) because we saw no whales yesterday. Commenced scraping the gum from our bone to prepare it for the market. The gum is a white & tender substance much resembling the soft part of a horse hoof.

Wednesday June 14th –- Calm & foggy. Cleared away after dinner long enough to lower for what proved to be a Sulphur bottom.

Thursday June 15th –- Weather mixed. Clear & foggy. Calm & blowing. We are now on the ground where we got our first whales & are cruising for more. Seen whales twice but not fair to lower. Spoke the *Condor*, a ship that was in port with us. As yet she has not got but one whale. A dead whale close alongside but too far gone to save. Another sail to leeward.

Friday June 16th –- Weather blowing & foggy. Nothing doing. All hands below reading, sewing & sleeping.

Saturday June 17th –- Rough in the morning. Tonight calm & smooth. Lowered this PM. The bow boat went on but missed.

Sunday June 18th –- Light winds & weather mostly clear. Lowered this AM and in a short time carried our boat steerer fairly into a large whale but the booby missed. After three hours more hard pulling, the larboard boat fastened & killed a goodly fish & by two o'clock we had him alongside. We have lowered twice since. Fastened once & drew (The larboard boat). Commenced cutting in.

Monday June 19th — Foggy this AM. Clear this PM. Finished cutting in before noon. After dinner lowered and our boat fastened for the first time to a loose whale but the iron broke. Tried two or three more but could not get on. We are now trying as fast as we can.

Tuesday June 20th — Rain last night. Today calm & fair. Lowered once but could not get on. Sail three miles off. Boiling & whaling.

Wednesday June 21st — Calm & pleasant & by far the warmest day we have seen in this Latitude. Lowered this morn but after three hours pulling came on board again, the whales having seen us. The Captain gave the mate a severe scolding for not doing exactly as he was told. Whether right or wrong the mate is a capital right whaler & the Captain knows nothing about it. For this reason the Mate takes the liberty to exercise his own judgment occasionally. Lowered again after dinner. Pulled all the afternoon. After several "pools" of whales & just at Sunset the bow boat fastened but drew. The sail we saw yesterday is cutting in today. Finished trying out this eve Lat. 51° N Lon. 1449 W.

Thursday June 22nd — Since morning the weather went from still & smooth has been growing wetter & rougher until now it is wet & rough enough. Lowered three times, once just at sunrise. The last time the larboard boat went on but missed. Have broken out & commenced stowing down in the ship's fore peak.

Friday June 23rd — This morn the wind died away to a calm but it proved that old Neptune was only rigging a new bellows pole for in less than two hours after breakfast, we had to close reef the M T sail &

take in the fore sail in which situation we now are. Nothing doing. All hands below, some too lazy to turn out to get their meals. On such a day as this friends at home are doubtless pitying the condition of the poor sailor. But tell them their pity would come more opportune upon a pleasant day for then we are hard at work but now doing nothing.

Saturday 24th — Wind light but sea rough. Finished stowing down. Raised whales three times. Prepared to lower for what proved to be fin back of the stewards raising. Our steward is aloft much & has raised many whales but as luck will have it we have taken none of them. The crew begin to believe that it is impossible to catch a thing of his raising.

Sunday June 25th — Weather pleasant & wind light this AM. Tonight it rains & the wind freshens. Lowered at 8 o'clock and after some maneuvering, the waist boat captured a forty barreler calf & by 3 o'clock had him in the blubber rooms. Lowered again at five & in an hour the bow boat had taken & turned up a hundred & fifty barrel whale. When struck he ran like lightning a short distance & breached his whole length out of water within a short distance of us. When I say that a boat has taken a whale I do not mean that that boat alone killed him, for three or four boats always fasten as soon as possible & have a hand in killing him. Mr. Smalley generally takes their life.

Monday June 26th — Last night the wind freshened so much that to make all sure we put a large hawser upon our whale. At 12 o'clock it blew a screamer & a heavy surge parted the fluke chain. The whale lay two hours by the hawser when that parted & let him adrift. At noon the weather had moderated so much that we made sail & since have been beating to windward in search of the lost, but it has been too foggy to raise him. Commenced trying in our calf at noon Tonight wind moderated with a high sea. Weather looks like more wind.

The Log Book

Tuesday June 27th — Blowing hard with a heavy sea. Running NW in search of better weather. Finished trying our calf & instead of making 40 barrels he will make hardly 35. The Captain had his legs jammed by a cask but not badly. Passed two whales.

Wednesday June 28th — Wind W & some milder than yesterday. Sea not so rough. Thick nearly all day. Running N in search of better weather.

Thursday June 29th — Weather pleasanter then it has been for some time. Very little wind & a bright sun. Still running N Westerly with R G Sails out. Finbacks all round. Three sails in sight.

Friday June 30th — Fresh N wind. Rain this morn. Pleasant this PM. Spoke this morn the ship *Amethyst* of New Bedford 8 mos. & 1400 barrels. Lowered at 7 PM & bow boat sank a 90 barrel whale. Hove to tonight to look for our lost tomorrow.

Saturday July 1st — This morn weather fair & clear. Lowered at 4 AM and after two hours pull came on board. No sooner than the boats were on the cranes, they were lowered again. The waist being first, fastened to one of two whales in the act of finning (a very large one). Three boats were soon fast but were unable to kill him until the fog became so thick that it was impossible to see a ships length.

Two boats then cut & with the Captain endeavored to find their way to the ship while we held on to save all our line which we did before running 1/4 of a mile. We then laid our course for the ship according to the best of our judgment. As good luck would have it, we were alongside in less than 15 minutes. We then commenced firing our cannon for the other boats & in less than an hour all were on board. Before they arrived however many were the plans laid as to our probable course if the boats were lost.

In an hour it was all clear again & we in the boats, there being whales on all sides. Came on board again to dinner. Lowered again immediately after & fore waist boat fastened again. But when she fastened, she was 10 miles away from the ship and alone with the ugliest fish we have yet seen.

When we fastened all that could be seen of us was our boat sail & the whale's white water. The ship & other boats made all the haste to reach us that human strength or canvas rags were able to exert well knowing our dangerous situation. Before any help had reached us the line had parted & the vile monster was at liberty. Two sails, two carcasses & our blasted whale in sight.

Sunday July 2nd –- Wind light & weather fair. This morn while looking for game we raised the whale we killed the night before last, brought him alongside & now have him in the blubber room. He is a larger one than we thought him. Lowered twice in vain. Sail to windward with a whale. Just at night a very thick fog set in with a fresh breeze & the Captain, in consideration of the bad weather & our fatigue, permits all hands to turn in instead of starting the works as usual.

Monday July 3rd –-Weather fair, wind light. Lowered once for a gallied whale. Trying out & making a bread pen in order to empty casks. Two sail in sight boiling. One hauled back his main yard to leeward, a signs for us to run down to him but it would not take.

Tuesday July 4th –- Weather fair but it has a bad look for the morrow. Our watch was called on deck before time to lower for a Sulphur bottom. At noon the boats (barring ours our slewer being lame) lowered. The larboard fastened & Mr. Smally was thrown overboard by a slight touch of the flukes under the boat, but without injury to himself or boat. At 2 o'clock we had the whale fast by a fluke chain. At half past 4 commenced & now at 11 o'clock there are but two blanket pieces overboard. In consideration of its being the 4th of July & of our getting a two hundred barrel whale the, the Captain has spliced the main brace twice & the grog has the credit of running in the pieces. All the forecastle save myself & another little Connecticut chap took their glass. The works stopped & all hands in tonight.

Wednesday July 5th — Weather same with occasional sprinkling of rain. Finished cutting & commenced trying. Two sail in sight. Whales all round but we are blubber logged & our deck covered with oil.

Thursday July 6th — Weather same. Commenced breaking out & stowing down the fore hold. Two of the slewers being sick, I have to take the place of one of them & stand before the works. This eve two sail ran down to us & the Captains of both came on board us & stopped till 11 o'clock. (The *Factor* of Hudson 24 mos. out & 1500 barrels & the *Envoy* 3100 barrels & 25 mos. out).

Friday July 7th — Weather pleasant. Trying & stowing down. Finished boiling this eve. For four watches I have assisted before the works & in two I have stood alone in the third & fourth mates place. Seven sails in sight. The Captain is carrying all sail to get out of their way. After dark had a gam with the *Hope* of Bedford & the *Mogul* of New London, 9mos 1500 barrels. Two boats came on board us & being before the works I played the slewer for the time.

Saturday July 8th — Very pleasant with a fresh breeze. Stowing down the last of what we call our first thousand barrels. Plenty of whales but going quick to windward. Lowered once. Stowing down. I have the easiest billet on board having nothing to do but see to the venting of the casks while others have the hauling of them.

Sunday July 9th — Weather pleasant, wind light. Spoke Bark *America* of Bristol, RI, 21 mos. 1000 barrels. Immediately after, lowered & the waist boat captured a large whale. Although fast by two o'clock, it was night when he came alongside.

Monday July 10th --- Weather lowering. Fresh wind. Cut in & commenced boiling. Spoke Ship *Mogul* again; boiling.

Tuesday July 11th --- Weather thick. Occasional showers. Wind light. This being the first whale after the first thousand barrels, we have doughnuts every night fried in the try pots. Today I have been Cooper & succeeded so well that the Captain says I must pay for my trade.

Wednesday July 12th --- Weather thick. In a little clearing we raised and brought alongside a whale. He had been dead some time but we shall get nearly an hundred barrels from him besides ¾ of his bone. He had an iron in him marked SMZ. While cutting in a boat from the *Miscasset* of Sag Harbour came on board us.

Thursday July 13th --- Thick with short clearing. Boiling & stowing down. Sail in sight. Lay in all last night sick for want of sleep. The constant work has about used up the crew.

Friday July 14th --- Weather clear, wind light with a little rain. Stowing & boiling. Finished trying. Three sails but no whales in sight. Mr. Smalley wanted to hoist the colors today, being the first time that all hands were on deck, some being sick all the time. Last night one of the hands fainted in the blubber room fairly worn out by fatigue. This is the second occurrence of the kind.

Saturday July 15th --- Weather thick & thin, wet & dry. Piped on pretty strong this AM but has fallen off since. Finished stowing down the last 500 barrels fare & have once more cleaned decks. Finbacks. Sail this eve.

Sunday July 16th — Weather fair, warm & tonight calm. Today for the first time since we took our first right whale I have washed & shaved & doubtless have improved my appearance. One sail & plenty of finbacks. The ship that came down to us last night was the *Maine* of Fair Haven, 250 barrels. Left home a little before us. Her Captain is in Oahu from injury given him by his steward in attempting to murder him. The steward afterward jumped overboard. From her we learned that the *Canada* from the same owners & sailed since is in search of us with orders. Likewise that we have done as well as any ship on the ground.

We are now looking for the *Canada*, running in the direction of where she was seen a fortnight since. The *Canada* is fitted to commence bay whaling on the N American coast & has a pilot with her for that purpose. It is possible that we may have to go with her if we find her. As this is a new scheme, I have a desire to have a hand in it. No ship has as yet tried it & if we go we shall be the first to lay open an entirely new ground for American enterprise.

Monday July 17th — Weather "Nasty", raining frequently. Lowered once and pulled several miles after a running whale. All hands breaking out between decks to be ready for more oil. Once more have our forge in operation. Spoke this eve the ship *Havre* of Havre 1000 barrels & a whale alongside.

Tuesday July 18th — Fair & foul. Frequent rain squalls. Lowered once in rain. Employed now in cleaning bone.

Wednesday July 19th — Light breeze & a little rain. Lowered twice. First time the larboard boat fastened but drew. Second time did nothing. The region we are now in seems to abound in small whales, sometimes 6 or 8 in sight at a time, none larger than 30 barrels. We are now running down the coast and within 50 miles of it. Expect tomorrow to be near the entrance to Queen Charlottes Sound into which we may run. Carrying all sail & standing whole watches tonight.

Thursday July 20th — Calm & pleasant. Lowered once but came home minus. Discovered a curious egg with which the water abounds here. It is about the size of a pea & looks like a human eye. Inside of what resembles the pupil is a yellow yolk no larger than a mustard seed.

Friday July 21st — Calm & pleasant. Scarcely breeze enough to ruffle the surface of the water. Sent a boat off at daylight to examine a dead whale. Reported rotten. Now are taking care of our bone by drying & stewing between decks. Humpbacks all round.

Saturday July 22nd — As yesterday. Lowered for a fast whale, too fast for us by half. Curing our bone.

Sunday July 23rd — Weather fair. Wind fresh. Running Eastward. We are further from land than I had supposed.

Monday July 24th — Weather squally. Wind strong. Spoke the *Milton* of New Bedford, 14 moths out, 2200 barrels. The *Dartmouth* of same in sight. Tonight came into a lot of whales. Lowered, bow boat fastened but it was two hours before he brought it to sufficiently for the other boats to fasten. Smalley soon did for him & by 11 o'clock we are below. Considering the time of lowering (6 o'clock) & the roughness of the weather we did very well.

Tuesday July 25th — Turned out at 3 o'clock & commenced cutting although it was rough & raining. In less than 15 minutes one of the guys parted & the Captain found that he was endangering the ship by continuing. Unhooked, made all fast & all hands turned in to a night rougher still. Clewed up M top sail. All the rest furled save M Spencer & F Topmast sail.

Wednesday July 26th — Turned out at 4 o'clock and commenced cutting although it was rough & raining hard. Being too rough to take in the head we let it go adrift. By ten we had all the blubber in & fair weather to boot. Since that time it has been almost calm with a little fog occasionally. Lowered twice but the rascals were too wide awake for us. Large ship five miles to leeward nearly all day. Now she is nearer in consequence of our running off.

Thursday July 27th — Weather fair & calm. Raised & sent a boat to look at a dead whale this morn. Reported unfavorable. The Captain was doubtful and went himself but left him behind.

Just at night we ran for a large ship. When within 2 miles, the stranger lowered & fastened to one in a gam. One of the whales came towards us & when within a few ship lengths lowered & gallied himself so thoroughly by running over him & by shouting that the poor fool brought to & lay rolling until the bow boat fastened. Although he did not run much yet he kept his flukes in such constant motion that the four boats could get but two or three poor lances at him. The bow boat finally got over his flukes & before she got off she was stove amidships & all of her stern knocked to splinters. Two of the crew took to the water & were picked up by us. The rest remained in the wreck until she capsized & were picked up by another boat in time to save all. Towed the skeleton aboard & made all snug. Until 11 o'clock the whale was snapping his flukes as fierce as ever, close too. We only hope to get hold of him tomorrow & pay him off.

Friday July 28th — Weather fair & wind light. Turned out at 3 o'clock. Made sail & commenced whaling. Since we have lowered four times, 3 times gallieing before we could fasten. The last time we fastened to a large cow with a calf by her but in putting in the second iron we cut our line & lost her. The sail that took a whale last night had him cut in before breakfast & was whaling again.

Saturday July 29th –– Weather fair, wind fresh. Commenced breaking out the main hold to stow our last fare below before breakfast. But we had hardly eaten when we lowered for a whale close to. Our boat was first & before pulling 6 strokes we fastened but after holding on some time the iron drew. Came on board at 10. Lowered again at 11 in a spot where the whales were so thick that we sometimes had several around us at once. After much maneuvering the Captain fastened and as soon as the mate could get a lance at him he killed him stone dead, hitting him in the heart, he at once took his life. Our little Jack steered the mate for the first time (the slewer being unwell) & he acquitted himself handsomely. Commenced cutting by 3 & by 7 he was onboard, Three sail in sight. One within a few miles. Boiling & whaling.

Sunday July 30th –– Weather fair, wind light. Lowered twice this AM for finbacks, mistaking them for right whales. At noon commenced breaking out the starboard after hold to stow our oil, but at 3 lowered for a right whale and by dark we had his tail in a fluke chain. Bow boat struck him & the Captain killed him. Boiling our yesterday's whale. Breaking out & catching whales gives us all the work we could ask for.

Monday July 31st –– Weather warm. Cloudy with a trifle of rain. Turned out at two & by one had our whale cut in. We think he will make over 150 barrels. Commenced stowing down our oil. Sails all round close to & far off. Some boiling & some not. The Captain says we have now 1500 barrels aboard.

Tuesday August 1st –– Weather fair. Wind fresh. Boiling & stowing down & between both the decks are completely covered with casks of oil & provisions. Spoke ship *America* of Stonington again with 1100 barrels.

Wednesday August 2nd — Wet & bad this AM. Since noon fair. Boiling & stowing. Stowed off the after hold, within 200 barrels. Lowered once for a whale that out ran us to Windward. Seen others all bound to the Eastward where the wind lies. Running NE, only one sail.

Thursday August 3rd — Cloudy & rainy at times. Finished boiling & commenced stowing oil in the larboard main hold. Seen whales going quick to windward. One sail.

Friday August 4th — Wind fresh. Weather thick & thin. Stowing down. Lowered once but the fog drove us back again.

Saturday August 5th — Wind light. Weather fair. Lowered at 3 AM, fastened and drew. Mr. Smalley fastened to another & parted, fastened again & had his line taken out; two other boats fastened after & after lancing him, both cut thinking the whale was sinking. The whale soon rose & by picking up our lines again we finally turned him up. By 11 we had breakfast & by 4 cut in. Call him 40 barrels.

Sunday August 6th — Weather clear. Wind fresh, so fresh that we double reefed the topsail at 3 o'clock. Seen finbacks & Sulphur bottoms. One of the latter turned flukes close to. The upper part of them was black the under white. Doing nothing. Waiting for another whale. Start the works tonight.

Monday August 7th — Weather fair wind strong. Running Southward. One sail & plenty of finbacks. Trying & scraping bone. We think it hard to be kept on deck to scrape bone.

Tuesday August 8th — Weather hazy. Wind light. Finished trying at noon. Nothing but finbacks.

Wednesday August 9th — Clear & calm. Warmer than for some time. Finbacks & bone scraping! More ships & are standing S SW.

Thursday August 9th — At 12 last night the wind sprung up from the SE & continued freshening until 9 AM when we took in everything save close reefed M. T. Sail & fore sail & lay to. Before we had been running SSW & as we think & hope, leaving the NW for the season for a warmer clime. This PM. the wind died away to a light breeze but made no sail, the weather looking bad.

Friday August 11th — Double reefed topsails. Breeze, rainy this AM. More ships. At sunrise pursued & spoke a Frenchman that has taken 1700 barrels this season. Learned that the *Amethyst* has filled up and gone home. Today we have been running eastward instead of southward as we had hoped. Finished scraping our bone. We are in a quandary as to what will be our next standing job. Wind S.

Saturday August 13th — Fair fresh wind from S. Running E NE under short sail. Standing whole watches tonight.

Sunday August 14th — Weather fine & fair. Wind light from NE. Steering E. Not a spout or sail in sight, something unusual in this quarter.

Monday August 14th — Weather fair. Wind fresh. Running E, wind NE, tonight squared yards & are standing S SE, this with the variation of the compass makes our course S. Plenty of finbacks.

Tuesday August 15th — Weather as yesterday. Stowed our last fare in the M hold this PM which makes it nearly full. Running SW with wind NW. Finbacks. Our library is at last open & with the exception of a few, there are none to interest me. Most of them are for pious old Methodists, the rest for children.

Wednesday August 16th — As yesterday. Nothing to do.

Thursday August 17th — Cloudy with fresh breeze. Lowered immediately after breakfast & except a few minutes for dinner we have been in the boats ever since, without even a dart. Took in sail & made preparations for a blow of which there is every appearance. Have seen three breaches & that they say, is a sure sign of foul weather. We had hoped never to see another right whale, but alas for us we are among them again near where we took our first right whales.

Friday August 18th — Weather as yesterday. Lowered three times, darted once & that's all.

Saturday August 19th — Some rain this morn. Fairer this PM. Lowered 4 times. Went on to several whales. Fastened three times & drew. Captain twice. Bow boat once. Captain slightly stove both times. The whales are so wild that it is next to impossible to get them.

Sunday August 20th — Weather fair, wind light. Lowered four times and fastened once by larboard boat but being scooping he was too deep to hold. The animal Krill, upon which the right whale feeds, is now coming up to spawn. The water is thick with them.

Monday August 21st — Weather cloudy with fresh breeze, lowered 3 times. Last time larboard boat fastened to a very large whale and before we could get fast, he started to windward, the rest trying to follow. By the time that we had pulled five miles the boat & whale were out of sight & remained so for nearly an hour when he got loose, not however until he had broken three irons, lost a lance & cut his small very much with his spade. We were fearful for a time that they were stove & the crew perishing. When the boat came on board the crew were wet through & shivering with the cold.

Tuesday August 22nd — This morn thick & blowing. Cleared away at noon. Made sail. Lowered at two, fastened to a large cow (larboard boat) & in lancing her we rec'd a blow from her fin that stove our bows & sent us onboard of her stern. Now at 10 o'clock we have her fast after having towed her 2 hours. She was so far to windward that the boats fairly drove her to leeward, one boat on each side guiding her with their lances.

Wednesday August 23rd — Weather fair & almost calm. Cut in our whale & commenced trying. Lowered once but not getting fast immediately came on board. The whales are thick around us. While cutting a large whale rose alongside ours and within 20 feet of the ship.

Thursday August 24th — Weather fair, wind light. Lowered at 9 o'clock on board at 11 o'clock having sunk a 75 barrel whale. Lowered again this PM. Fastened (larboard boat) but iron broke. Trying.

Friday August 25th — Calm & pleasant. Lowered three times. Last time our boat fastened to a large whale while feeding. We had fair expectations of getting him until he got in company with a loose whale. We came near being knocked into chips and the larboard boat, as is her want, getting into the midst of them had a hole stove in her bottom that disabled her by the loose whales coming up under her stern foremost. We cut line, picked them up & brought them on board. The other boats getting loose, we cast our fish. Trying still.

Saturday August 26th — Weather fair with a light breeze. Lowered twice in vain. Stowed down enough oil to check off the lower main hold & broke out to stow the fore peak. Finished trying.

Sunday August 27th — Mostly thick. In one light spell saw whales to windward.

Monday August 28th –- Weather thick & blowing. Stowed down our oil in the fore hold which almost fills it. This whale makes us over 1500 barrels. Blowing so tonight that we close reefed M T sail & took in fore sail. Two of our boys had a square set to tonight but at the second round one of them got his nose scratched and gave in. They frightened one of our Kanakas so badly that he ran aft & told the Captain that someone had struck him & the Captain immediately came forward & threatened the offender with a flogging if what had not been done were done again.

Tuesday August 29th –- Still foggy & blowing. Nothing done save repair the larboard boat.

Wednesday August 30th –- Weather same. All hands below reading, mending & sleeping. Spoken by the *Golconda* of New Bedford 26 mos. out, 2300 barrels. 800 Sperm, 1500 Right. Her Captain came on board although it was so rough that anything but a whale boat must have swamped.

Thursday August 31st –- Weather fairer than yesterday. Carrying topsails. Whales all round us. Tonight took in sail early to be in readiness for the morrow. Scraped our bone.

Friday September 1st –- Weather somewhat windy with a heavy sea. Had a bit of rain squall. Lowered at noon and in spite of our efforts to the contrary, our boat fastened & away we went to windward, the boat actually leaping from one sea to another, taking in water so fast that it kept me constantly bailing. After he had run us about 10 miles Mr. Sanford cut the line attempting to spade him. When we got on board we were a miserable set, every thread wet and nothing to pay for it.

Saturday September 2nd –- Pleasant & calm. Lowered this AM & sunk a 100 barrel whale by too much lancing. The *Golconda* close to boiling. She lowered at the same time with us and for the same whale but neither fastened. This eve the Captain is on board the *Golconda* and her mate and a boats crew with us having what we call a gam.

Sunday September 3rd –- This day has proved an eventful one to us. Last night at 12 a breeze sprang up which has at length become a close reefed top sail gale. Lowered at 7 AM with a fresh breeze. Our boat fastened & saved a small whale, sending another off spouting blood. By the time we got him alongside it was quite rough. Before we got the first piece in, the fluke chain parted but the whale held by the fin until we could get a larger chain upon him. Commenced cutting again but soon found that it was impracticable. The Captain got mad cursed all right whales. Cut away the whale and put all the sail out the craft can stand and is now booming it for a warmer clime. Steering SSW with wind on the larboard quarter. We soon hope to see a port, probably Oahu. Among other things the Captain has got a dog about as large as a "pint of cider".

Monday September 4th –- Wind died away last night until she carried whole topsails. Since noon however, the wind has gradually increased until at ten this eve she labours under nothing but close roofed MT sail & stay sail with Royal yard sent down. Latterly in reefing the topsails, I have generally taken one of the bearings from the boat steerers. This for a green hand is considered by them as a supererregation, it being their place.

Tuesday September 5th –- Pleasant & calm with heavy sea. Lowered twice, fastened once, but drew (larboard boat). Boiled out the blubber that we saved from the last whale. Got about five barrels. This eve, light breeze from N. Course S.

Wednesday September 6th — Weather thick with all the wind we can carry. M T G sail too. Took in the fore & Mizzen at 4 o'clock. Running S tacks, wind abeam. Set the M sail for the first time today.

Thursday September 7th — Last night the wind continued freshening and we were taking in sail until 7 o'clock AM when we had nothing out but close reefed top sails & fore sail. From 10 last eve till 10 AM it rained sufficiently to make us uncomfortable. As soon as it cleared we commenced washing the ship with a strong lye made from the cinders taken from the works while trying. We have done much today but shall have another job the next fair day. This eve we are running SE by S. Close hauled S tacks & same sail as at morn. While it blew & rained the hardest this morn we turned up the bow & waist boats to save them. The sea has been so rough that once she pitched all her forward spars well under water, a thing that has not been done before on this voyage. For some time past I have been afflicted with an old complaint, a deep seated pain in my eyes occasioned by reading too much, although I have not read two hours per diem this three months. I fear I shall never forget studying Greek by candle light.

Friday September 8th — Still blowing. Still a heavy sea. F T sails close reefed. Occasional squall. In one we took in F & Mizzen T sails but set them soon after.

Saturday September 9th — Occasionally thick & wet. Wind lighter. Shook a reef out topsail. Set the sail & jib. Still washing ship. Making SE.

Sunday September 10th — Weather gradually growing better & wind lighter. Now carry M Top G sail. Steering S SE close to the wind.

Monday September 11th — Light winds & squalls. Wind SW running close. Lat. 40° N.

Tuesday September 12th –- Weather squally. When not in a squall the wind is South and light (close hauled). Still cleaning up things dirty. The weather is so much milder here than on the NW that I have discarded woolens & shoes and adopted ducks & barefoot. A large school of (?) followed us some time.

Wednesday September 13th –- Wind light & steady from S SE running close. Washing spars & repairing an old boat. One breach.

Thursday September 14th - Pleasant & warm. Wind light making five knots. Steering S close. Washed ship outside. Nearly through with cleaning her. Commenced a letter home.

Friday September 15th –- Wind light and pleasant save a little rain. Steering SW. Still polishing.

Saturday September 16th –- Weather mostly pleasant. A little rain. Tarring rigging & fixing small matters. Steering SW, wind light & aft, set on F T sail this morn.

Sunday September 17th –- Weather as yesterday. Wind aft driving us 9 knots. I think we have taken the NE Trades which take us directly to Maui. Steering SW by S.

Monday September 18th –- Weather as yesterday. Some rain, wind & course same. Sail that was in sight last night is so still but are nearing her. Several breaches. Work is cleaning small affairs & patching rigging.

Tuesday September 19th –- Wind same. Occasional squalls. Course SW 1/2 S. The wind took our studding sail nearly out of the bolt ropes last night. Bent a new one today also a new M T G sail. Commenced bunching our bone. The sail that was ahead of us last night is now far astern. Writing to Fish. Flying fish again.

Wednesday September 20th –- Weather fair. Wind in same quarter but growing gentler. Course same. We are now bunching our bone to send it home if we have an opportunity. We are now about 5º from land.

Thursday September 21st –- Weather fair with little rain, steering SW. Wind on larboard quarter. Still at our bone. Expect to make land tomorrow.

Friday September 22nd –- Weather fine. Wind light. Course SW 1/2 W. Square yards. Not yet raised the land but suppose it is in sight. Finished our bone & have been putting our ship in trim for getting our ground tackle out. Expect to use it tomorrow.

Saturday September 23rd –- Weather fair. Raised land at daylight having laid back 4 hours. After getting everything ready & working in all ways with light breezes we at about 4 o'clock dropped our anchor among a fleet of 20 sail. We find that we have done as well as the average of NW Cruisers. Sent a raft of casks ashore for water.

Sunday September 24th –- Weather fine. After giving the decks a thorough scrubbing, our watch was told to get ready for shore. As soon as I landed I laid my course for the Mission houses upon the side of the mountain in order to see how they live. In one I saw a handsome American girl (without corsets) & it is so long since I have seen one of the genus that it almost shook my resolution with regard to the sex. When returning from the mountain I was followed by a Kanaka who, from his actions, had an intention of arresting me as a runaway. Since we were here before they have raised a bit of an army for the purpose of keeping the peace. I should judge them all to be officers from their uniforms, blue with white stripes & epaulettes & Oh! How crank! 3 sail came in today.

Monday September 25th — Weather fine. All hands aboard scraping, painting, breaking out shop & sending our bone on board the *John* of New Bedford, bound home.

Tuesday September 26th — Weather fair. Larboard watch on liberty. Ours painting spars. Three sail came in and one went out. I already wish myself again at sea. I cannot eat fresh grub and but a little fruit.

Wednesday September 27th — Weather same. Our watch on shore. Same story. Sent off to another ship some meat & bread, brought off wood & painted ship. Three ships sailed & three are coming in.

Thursday September 28th — Weather same. Larboard watch on shore. Ours getting off wood, cutting it up & myself & a boat steerer painting ship. Wet hold this morn for the first time. This job we expect to have at least once per week hereafter. It is done to prevent the calks from shrinking while in hot weather. We have some job or other to use up part of our liberty days.

Yesterday we had to heave up our anchor & let go the other because of the first ones being fouled. It had so many turns of the cable around the shank and flukes that it dragged with but a light breeze. This eve we had a genuine breakdown on board us. There were nearly fifty sailors on board us from other ships with a fiddle & tambourine player & while they stopped our forecastle deck was a dance hall. Our steward treated the gemman to a bucket of soitchel. One of the larboard watch has come off twice well intoxicated. Thus far he is the only one. The rest come off sober as yet.

Friday September 29th — Weather cloudy. Somewhat cool. Our watch on shore. Nothing new save that I found an old bullock catcher (an American) with his appurtenances.

Saturday September 30th --- Weather as yesterday. All hands on board breaking out & taking on board the *Nye* of New Bedford our sperm oil.

Sunday October 1st --- Weather fair. Larboard watch ashore. Starboard watch pulling the officers around nearly all day.

Monday October 2nd --- Weather same. This morn before going ashore we hove up another fouled anchor & hauled nearer in shore by getting some sail & towing. Nothing new.

Tuesday October 3rd --- Weather same. Larboard watch ashore & we stowing the raft we brought off this morn. Our 4th mate has asked for his discharge & is to get it no one objecting.

Wednesday October 4th --- Weather same. All hands on board breaking out shocks & stowing water that we got off this morn. Got wood, potatoes and onions.

Thursday October 5th --- Weather same. All hands aboard getting off our last water & fruit & preparing for sea. Our fourth mate left us today & the Captain has procured another from the *Marcia* homeward bound.

Friday October 6th --- Same as yesterday. All hands preparing the ship for sea by lashing everything loose & preparing sails.

Saturday October 7th --- Weather same. Naught doing. This PM six of us went ashore for the last time & one had to be hoisted on board again. I had an opportunity cultivating an acquaintance that I have made here. He is a young fellow from Central New York that came for his health & is the only one that I can find any comfort in associating with. He belongs to the *Copia* of New Bedford.

Sunday October 8th — Weather same. Although Sunday, yet we have to heave up our anchor & loose our top sails. This eve we are laying to for daylight to go ashore at Oahu.

Monday October 9th — This morn we were at 8 o'clock, near enough to lower the boat & it being our boat, I of course went nor did I return until sundown.

In the harbor were several ships, men of war, merchant-men & whalers. As soon as I landed I went to Mr. Damon's but, not finding him at home, returned to the boat expecting her to shove off soon but Mr. D soon found me & obtained permission for me to stop on shore until night. Spent the day at his home. He is very pleasantly situated with all the comforts and luxuries of life that he could have at home. His wife (a fine woman) has a fine collection of shells & curiosities, chiefly presents from seamen. He has an infant added to his family. I received some of papers & books.

Left the Captain ashore to come off in the morning. Brought off two men, one an old cook that has been forty years at sea. The other is a man that ran away from us at these Islands last spring. He came of his own accord.

Tuesday October 10th — Weather fair. This morn a boats crew of officers & slewers went for the Captain and returned at 10 o'clock & since, we have run south with the trades on the Larboard Quarter. Had our watch below this PM. Anchors & chains being stowed.

Wednesday October 11th — Weather fine. Wind variable. Today have run S by E with wind E by N. Commenced the manufacture of spun yarn again. Fixing a new boat for Mr. Smalley. Some of our crew are suffering the common evils of licentious habits.

Thursday October 12th — Weather same. Wind fresh & NE. Steering S SE. Placed the new boat on her cranes this PM.

Friday October 13th — Weather same. Wind strong. Put up T G sails this eve. Repairing all sails.

Saturday October 14th — Weather squally. Wind strong. Double reefed torrents all the time. We are to wet holds every Tuesday & Saturday.

Sunday October 15th — Weather unpleasant with frequent rain squalls. Shook the reef out this morn & set M T G sail but took it in this eve. Wind strong & E, steering S by E.

Monday October 16th — Weather squally & has forced us to double reef top sails & furl Main sail. Repairing sails. Bent a Mizzen Top Gallant sail this PM that we patched this AM.

Tuesday October 17th — Weather better. Less rain & wind. Carry T G sails. Steering S on the wind.

Wednesday October 18th — Weather mixed. Light squalls & little rain until just at sundown when we had a heavy squall & much rain. Took in T G sails & reefed topsails.

Thursday October 19th — Weather same. Carrying all sail. Steering S on the wind. Run off this morn in search of Fanning's Island which we are near. The Captain says there are so many reefs & rocks in this region that he is afraid to run at night. We have accordingly worn ship & are standing on the other tack. Setting up stays & back stays. Mr. Simmons has commenced repairing his stoven boat.

Friday October 20th –– Weather pleasant save some light squalls. Made Fanning's Island this morn. The land is so low that it can be seen but a short distance & is entirely covered with coconut trees. We were in hopes of getting ashore for some. These Islands lie 3 1/2 degrees north of the line & 149º W Lon. Steering S SW on the wind.

Saturday October 21st –– Weather wet with frequent squalls. Wind light most of the time but we have to clew up T G sails several times per day for squalls. Making sennit & spun yarn.

Sunday October 22nd –– Weather fair & pleasant. Wind light from S SE running close. Making South. This the first day for some time that has passed without squalls. This eve the Captain called all hands aft & divided us into three watches, two of which are to be constantly on deck through the day & one at night. Two of the watch are to stand aloft forward, one at the mizzen & three to take the wheel.

Monday October 23rd –– Weather fine. Wind light. Course on wind. This is the first day of our cruise. Tonight took in T G sails & double reefed Topsails.

Thursday October 24th –– Weather fine. Wind lighter. Repairing sails. Shifted foresails. This PM I raised what was thought to be a sperm whale breach.

Wednesday October 25th –– Weather same. Wind same. Tacked ship last night & steering NE on the wind. Repairing sails. Shifted fore top & T. G. sails & M. T. G. sails. Blackfish.

Thursday October 26th –– Weather & wind same. Being S of the line & in the SE trades, we have a steady breeze & by tacking ship occasionally, we keep in the same latitude. Several finbacks.

The Log Book

Friday October 27th –- Weather & wind same. Working the ship on all tacks but have only smelt a whale. Repairing sails. Shifted M. top sail.

Saturday October 28th –- Weather fair. Sailing on a S tack--free. Finbacks--. Wind fresh. One of our number has gone. A young fellow from Connecticut fell from the Mizzen Top Gallant crosstrees, struck on a line tub, hurricane house & the chin of a water pipe. His head was badly cut, his arm broken & hip badly jammed. He lived but three hours. We launched him overboard after a chapter & prayer by the mate. Lat. 2° 1' S. + Lon. 133° 12' West.

Sunday October 29th –- Weather & wind same. Course NW, nothing done or seen.

Monday October 30th –- Weather same. Wind fresh. Today we caught a sperm whale from a school. Very small. Bow boat fastened. This PM cut him in. Are to try tomorrow.

Tuesday October 31st –- Weather, wind & course same. Lowered but the school were too quick. Commenced trying this eve.

Wed November 1st –- Weather fair, almost calm. Finished trying our fare of 16 barrels. The water here is almost brown with sharks, each large enough to take care of a man. The Captain & mate have had another blow up. The mate says that unless he can be mate, he will be nothing. Now that we are sperm whaling the Captain takes every occasion to abuse him & he says that he will stand no more of it.

Thursday November 2nd –- Weather same. Wind light. Repairing sails.

Friday November 3rd –- Weather & wind same. Saw a sail. Blackfish & porpoises. Repaired & bent again the spanker.

Saturday November 4th –- Weather & wind same. Set our fore rigging. Another sail just ahead of us for which we are running. Have not as yet taken in sail.

Sunday November 5th –- Weather hot & almost calm. The ship we ran for & had a gam with last night, was the *Levant* of Marcham, 13 months out & 200 barrels of Sperm. Had another gam with them this PM. Steering W. Blackfish & porpoises.

Monday November 6th –- Weather as yesterday. Nothing save setting up Main Rigging.

Tuesday November 7th –- Weather same. Steering Northerly. Gaming with the *Levant* again which we raised this AM. Finbacks & porpoises. Set up Mizzen rigging.

Wednesday November 8th –- Weather same. Finbacks, grampuses & porpoises. Steering S SW. We are now about out of work & to keep us busy, the anchors & bobstays have been pecked to clear them of rust. Likewise reserving backstays.

Thursday November 9th –- Weather same. Serving Backstays & setting up F. T. rigging. Lowered at 3 PM for a large whale of my raising but could not get on.

Friday November 10th –- Weather same, tonight calm. Raised nothing. At 4 PM lowered for nothing. Beat all the other boats for nothing. Came on board with nothing save weary limbs. Caught a large shark.

Saturday November 11th — Weather same. Out of work, nothing to do but make sennit. Saw a large school of Algerines, Blackfish & porpoises.

Sunday November 12th — Weather same. Blackfish & porpoises.

Monday November 13th — Weather same. Lowered this AM for a school of whales without fastening. This PM lowered again & the larboard boat again missed.

Tuesday November 14th — Weather same. Lowered at Noon for a school. The starboard & larboard boats fastened but drew.

Wednesday November 15th — Weather same. Lowered this morn with better success than before. When we came on board at 4 PM we brought 3 whales with us. Small cows from a school. This eve we have one & a half cut in. The whales were got by the larboard, waist & bow boats. Ours struck first.

Thursday November 16th — Weather same. This morn finished cutting in before breakfast & started the works soon after. Wind E NE, standing N. Last night in my watch on deck I was employed with a boat spade upon the numerous sharks that were tearing off the blubber.

Friday November 17th — Weather same. Trying our blubber & trying in vain to raise more. Lat. 15° Lon 170° W.

Saturday November 18th — Weather same. Wind fresher. Finished trying last night & today cleaned up the decks. We got about 50 barrels from the three fish.

Sunday November 19th — Weather fair save for an occasional squall.

Monday November 20th –- Weather fair. Wind strong. Ran all day under double reefed topsails. Had some hard squalls last night. To keep us at work the mate has given us a lot of scrub teeth to make. He is hard pushed to find employment for all hands.

Tuesday November 21st –- Weather fair. Wind light. Broke out the starboard after hold & stored our oil today. I have seen nothing to remind me of home so much as did a lot of ballast stone that we hoisted from the hold. It was genuine Mica Slate from the hills of New England. I could almost see a plough among them & I believe I did see a hill of potatoes.

With the 18th of this month I commenced writing letters to all of my male friends at home, taking their names as chance gave them to me. I have done almost all kinds of work for myself since leaving port. I have repaired all my clothes, made a pair of duck pants (tarry enough), a good Sou'wester & a scrubbing brush. Our mate advises me to commence navigation but I say not until my letters are all written which will use up my spare time for a month to come.

Wednesday November 22nd –- Weather fair. Light breeze in AM. Calm & cats paws this PM. Blackfish. Finished coopering & stewing the meat we broke out yesterday.

Thursday November 23rd –- As yesterday. Little doing.

Friday November 24th –- Weather fair in AM & wind light. In PM, fresh & squally. Took in T G sails & this eve reefed Top sails.

Saturday November 25th –- Weather squally with light winds. School of large blackfish.

Sunday November 26th –- Weather fair & wind fresh.

Monday November 27th –- Weather fair & wind fresh. Nothing new save selling the clothes of the man that we lost a while since.

Tuesday November 28th –-- Weather fair. Wind fresher. Saw a bone shark the, largest gill fish & a breach. Shifted mainsails & repaired jib.

Wednesday November 29th –- Weather fair. Wind strong & sea rough. No T G sails. Nothing to do but to make sennit & catch Bonita. The officers amuse themselves after four PM in shooting boobies. I think I could beat the lot of them.

Thursday November 30th –- Weather squally. Wind light. Nothing new.

Friday December 1st –- Weather fair save very light squalls. Saw several carcasses. Suppose them to be our own that we left a long way from here. There is here a strong westerly current that sweeps everything to the westward. The Captain has been looking for a reef that has been seen but once or twice. His intention was to survey it & fix its place upon the chart.

Saturday December 2nd –- Weather fair by day squally at night. One spout.

Sunday December 3rd –- All as yesterday.

Monday December 4th –- All as yesterday save that we are keeping off to the westward.

Tuesday December 5th –- Weather fair. Wind light. Lowered this AM but the school were too fast.

Wednesday December 6th –- Weather fair & calm. Commenced making mats for Harness casks & scuttle butt. A breach. Caught two Dolphin & had an opportunity of seeing them change colors while dying.

Thursday December 7th –-Weather as yesterday. Lowered and took a large Blackfish this morn to the starboard boat. The fourth mate headed the boat & his bowsman steered him. Four times he was carried & four times missed. The first mate then fastened himself.

Friday December 8th –- Weather fair with light wind. Raised breaches just.

Saturday December 9th –- Weather fair. Wind stronger.

Sunday December 10th –- Weather fair, Wind light. Steering SW before the wind.

**NOTE: Evidentially a day has been lost here.

Monday November 12th –- Weather fair. Wind light. Steering W. Making mats still. The officers have about given up trying to keep us at work. We have fairly worked them out. Half the time we have nothing to do.

Tuesday December 13th –- Weather wet with frequent squalls. Took in sail NW fashion tonight. Nothing out but double reefed M. T. sail & foresail, staysails & spencer.

Wednesday December 14th –- Weather squally. Wind increasing. Ran this AM with reefed top sails. This PM took in another reef & Jib. White water from something.

Thursday December 15th –- Weather badly. Wind sometimes strong & again light. Last night had to take in the F & Mizzen top sails for squalls. Finbacks.

Friday December 16th –- Weather badly & squalls. We carry all sail day & night. Steering E by N. Blackfish & Jumpers.

Saturday December 17th –- Weather better. Wind strong. Set T Mast Studsail to get along the faster. Blackfish & porpoises. Have another litter of pigs to attract attention.

Sunday December 18th –- Weather curious. Fair & squally by turns. Just after supper all hands were called to brace round (the ship being taking aback) & take in sail.   From M. T. G. sail we have reduced her to close reefed topsails. The wind has, since noon, shifted to almost all points of the compass. It now is west as of this morn. Steering E by N.

Monday December 19th –- Weather fairer. Squalls light. Steering E by N. Wind N. Commenced making a new M T G sail. Bent one in the place of the one that blew away last night

Tuesday December 20th –- Weather fair. Wind moderate & fair. Course same. Set TG sail. Saw Blackfish.

Wednesday December 21st –- Weather lowering in AM. This PM a succession of rain squalls during which the wind shifted considerable. Warned by this, we double reefed Top sails & were hardly done when the old squall struck us bringing the wind aft. Steering NE. Before night we close reefed & thus remain. The Captain says we are near some Islands that give us this weather.

Thursday December 22nd –- Weather squally. Have had to clew down topsails for squalls. Spoke the ship *Rosalie* of Warren 17 mos.

700 barrels Spoke her again this PM & her Captain came on board. Took in sail this eve.

Friday December 23rd –– Weather since breakfast fair. Carried all sail. This eve took it in. The *Rosalie* in sight all day. Set up back & bobstays.

Saturday December 24th –– Weather is good. Wind NW. Kept her off before the wind this morn & before breakfast raised a school of whales & before noon had four of them fast by the flukes. One to the larboard boat, one to the bow boat & two to the waist boat. Cut them in this PM. Were spoken by the bark *Damon* of Newport 12 mos. 300 barrels

Sunday December 25th –– Weather fair. Wind light. Boiling & gaming this eve with the *Damon* & *Rosalie*.

Monday December 26th –– Weather same. Finished boiling & got about 50 barrels. Since noon the Captain & his boats crew have been aboard the *Rosalie* & her mate with us. The Captain has given 3 barrels of flour for a spy glass, both of ours being spoiled.

Tuesday December 27th –– Weather fair. Wind light. Cleaning up. Finbacks & two sail.

Wednesday December 28th –– Weather a little squally. Wind light. Finbacks & Breaches. One sail. Bent a new M T G sail.

Thursday December 29th –– Weather fair. Wind light. Passed the *Rosalie* this morn. Lowered at noon for a school of whales & gallied them. Caught 500 wt. of Albacore & salted them for port.

Saturday December 30th –– Weather squally. Wind light. Caught a shark. Killed a hog. Saw a water spout about 3 miles off.

The Log Book

Monday January 1st, 1844 -- Weather as yesterday. Sail close to us sailing as us NE. Another year has passed & passed at sea but it is my hope that another New Year's Day may find me by my father's fireside.

Tuesday January 2nd -- Weather squally. Wind light. Spoke the ship *Three Brothers* of Nantucket 29 mos. 1600 barrels. Likewise the *California* of New Bedford 19 mos. 1400 barrels. Since noon the three Captains were with us the mates in the *Three Brothers* & the second mates in the *California*. The Captain discovered some new islands two days since & all three are running for them. Lat. 1° 50' S Lon. 171° W.

Wednesday January 3rd -- Weather fair. Wind light. Beaten in sailing by both vessels. The Captain is on board the *California* since noon.

Thursday January 4th -- Weather squally. Wind moderate in AM. Strong this PM. Both sail near us. Bent a new M sail that we might try our speed with the *California*. Then just abreast of her the pig that we got from her last night jumped overboard & lost us the race saving him.

Friday January 5th -- Weather fair. Wind fresh. This PM the Captain is on board the *Three Brothers*. The *California* is just in sight to windward. Why we keep together is to us a quandary. We are working to the Northward but where for we know not. We are much in doubt about our next cruise. Some say it will be on Japan some on the NW & there we all wish it to be.

Saturday January 6th -- Weather squally. Wind fresh. This PM we have been trading with the *Three Brothers*. For 3 bolts of canvas that we gave we received 4 casks of bread & 3 barrels of flour. Got a load of wood to boot. A sail to windward. Just heard that a Frenchmen on NW lost save one, all of 3 boats crews at a lowering.

Sunday January 7th — Weather fair. Wind light. The *Three Brothers* just to leeward of us.

Monday January 8th — Weather fair. Wind some slowed down. Our last fare of oil in the A hold. Another "gam" with the *Three Brothers*.

Tuesday January 9th — Weather rainy Wind variable. Ran a course today to find the Islands I mentioned but have not seen them yet. A boat from the *Three Brothers* came to us tonight.

Wednesday January 10th — Weather squally. Wind light. Nothing new save that we came near drifting into the *Three Brother*s last night. We had all sail out & a tow rope ready but a light breeze saved us.

Thursday January 11th — Weather fair. Wind fresh. Blackfish. We have finally lost sight of the *Three Brothers*. She has gone to the Westward to be ready for Japan.

Friday January 12th — Weather fine. Wind fresh. Naught but mat making.

Saturday January 13th — Weather fair & wind same. Our community has today met with a severe loss in the death of our old sow, the last of three swine that we brought from home. She died suddenly in consequence (as we suppose) of eating too much fish. The lady was launched this morn with due solemnity, She has left three small children to lament her loss besides a numerous circle of friends. I am sorry to state that the orphans have already forgotten their parent & appear happier without than with her.

Sunday January 14th — Weather fair. Wind as yesterday. Nothing else.

The Log Book

Monday January 15th –- Weather fine. Light breezes & calms. Last night we had a fair wind with rain. Our course, when we can lay it, is E by N. Scraped & varnished the bright work. Mat making as usual.

Tuesday January 16th –- Weather as yesterday. Nothing more.

Wednesday January 17th –- Weather squally. Wind variable. A ship close to us this morn, now far to leeward.

Thursday January 18th –- Calms & light winds with a thin drizzling rain. Boo.

Friday January 19th –- Weather fair. Wind fresh. Saw a finback breach. Scraped & painted the iron work. For some time we have carried all sail day & night in order to get to the Eastward but the current has carried us 1° to the Westward of where we were when we left the *California*.

Saturday January 20th –- Weather as yesterday. Lowered just at night for a lone whale but lost him in the dark. Took in T G sails & M sail.

Sunday January 21st –- Weather as before through the day. Last night we had rain in abundance. Wore ship twice & before morn hauled aback for fear of the old sand Islands. From 11 till 5 we steered west but have not yet seen the land.

Monday January 22nd –- Weather fair. Wind moderate. Spoke & gamed with the Ship *Mary Ann* of Fair Haven 10 Mos. 600 barrels

Tuesday January 23rd –- Weather fair & wind as yesterday. The *Mary Ann* went out of sight this AM.

Wednesday January 24th –- Weather same. Saw a breach this PM.

Thursday January 25th –- Weather squally. Wind variable. A breach & a sail.

Friday January 26th –- Weather fair. Wind light & fresh. Sail astern. Parted the luck of F T sail. Took it down & are making a Mizzen Top sail of it.

Saturday January 27th –- Weather as yesterday. Wind fresh. Sail to windward this morn. We have almost run out of wood & if we are not now running for port we must soon.

Sunday January 28th –- Weather fair. Wind fresh.

Monday January 29th –- Weather squally. Wind changeable. Had a squall last night that rent the fly jib. Bent a new one today. Also broke out all the shocks.

Tuesday January 30th –- Weather fair. Wind fresh. Commenced rattling down rigging.

Wednesday January 31st –- Weather as yesterday. Spoke ship *Foster* out of Nantucket 29 mos. 1000 barrels. Gaming with her. Rattling.

Thursday February 1st –- Weather, wind & work same.

Friday February 2nd –- Weather squally. Wind variable. Rattling.

Saturday February 3rd –- Weather as yesterday only worse. Rattling still while rain will permit. Thus far I have remained in the bunks I first took possession of, but today I have effected a move & I feel like a cat in a strange garret. If a landsman had seen the goods & chattels that were stored therein he would have doubted their ever returning again much less a place for my own corporation. We commenced standing whale watches to night & are sure now that we are bound for the S--h I--s. Our course is NNE but we can head no better than N NW in these trades.

Sunday February 4th –- Weather clear. Wind strong. Reefed topsails this AM. Shook them out again this PM & set the M T G Sail.

Monday February 5th –- Weather squally. In one we had to clew up & down & just after raised a whale from deck. Lowered three boats but saw him only once after. We head about N on the wind.

Tuesday February 6th –- Weather clear. We carry double reefed topsails through a heavy sea.

Wednesday February 7th –- Weather clear. Still under double reefed topsails. We are keeping a sharp lookout for land. What land I know not. Making N by W.

Thursday February 8th –- Weather same. Wind milder. Carried M T G sail this morn but are now with single reefed top sails. Sent down M T G sail & shortened it then sent it up again. Making N by W.

Friday February 9th –- Weather same. Carry double reefed T sails. Making N. NW.

Saturday February 10th –- Weather squally this PM but the wind has been sufficiently moderate to carry all sail. Making N by W.

Sunday February 11th –- Weather squally. Making N by E under double reefed top sails. Took in sail & wear ship this eve for fear of land ahead. We are now in Lat. 17° N & Lon. 179° 4' E.

Monday February 12th –- Weather same. Made all sail this morn but have seen no land although the weather feels as though we are to leeward of some piece of terra firma. Took in some sail this eve but have made it out again.

Tuesday February 13th – Weather fair. Wind good. Whole sail breeze. Let up some of the head rigging that we may carry the fly jib. Finbacks. Making N.

Wednesday February 14th – Since morn the wind has been freshening with some rain. Can just carry M T G sail. Split the jib from head to foot. Bent a new one in her place. Making NW.

Thursday February 15th -- Weather fair. Wind changeable & variable. We head from E to NW. Repaired Jib & staysail. Our old English cook has hove out with a lame leg that he may be discharged in port. Professedly a pattern of cleanliness he has turned in with more dirt & slush about him than all hands can show.

Friday February 16th –- Weather Fair. Wind assorted. Last night it came from the SE so that we could steer NE free. This morn it was S. Set the G T S sail. Since, it has hauled to SW where it remained until at 6 PM when it came in quantities from the NW. We are now under double reefed topsails on the wind making N E by N. Set up fore Topmast backstays & scraped stantions.

Saturday February 17th –- Weather cloudy & cold. Wind from NW to W keeping us on the wind under double reefed Topsails. This eve turned up the waist & bow boats. I expect we are to tack ship. Since we left the line it has been growing colder & now we have to adopt our heaviest clothing.

Sunday February 18th –- Weather fair. Wind light. Carrying all sail. Make NNW on S Tacks. We are now just to the Westward of Ocean Island where the Parker lost her mast two seasons since.

Monday February 19th –- Weather fair. Wind light this AM fresh this eve. It has hauled from SE to SW. Steering NE. Set M T G S sail. Set M T Back stays & Mizzen stays.

Tuesday February 20th –- Weather foul enough. Wind S Westerly. We commenced taking in sail last night at midnight & kept taking in until we had reduced her to close reefed Main topsail & foresail. To furl the fore & Mizzen our watch was turned up. Early in the morn we turned up the larboard boats. Took the bow on deck. Soon after turned up the starboard boat. Notwithstanding this precaution the waist boat is handsomely stove. Being before the wind we rolled both rails under, shipping a sea at every roll that washed light articles about at a fearful rate. Steering E NE this eve under reefed topsails & main sail. Wind on leeward beam.

Wednesday February 21st –- Weather clear. Wind gradually grows lighter. We have just set M T G sail over single reefed topsails. Course ENE near the wind on larboard tacks. Turned down the boats this morn. We had today the worst bread we have had on this voyage. It has been wormy the past six months but we could not grind it as it was quite too juicy.

Thursday February 22nd –- Weather clear. Wind fresh. Carry M. T. G. sail on the wind. Making SE. This eve tacked ship. Took in the waist boat to repair. Polishing stantions & spare spars.

Friday February 23rd –- Weather cloudy. Wind fresh. Tacked this AM & are making E SE on the wind on larboard tack. Repairing boat.

Saturday February 24th — Weather & wind same. Tacked ship this eve & making W NW. Lat. 30° N Lon. 171° E.

Sunday February 25th — Weather cloudy. Wind blowing. Running E under double reefed topsails. A heavy head sea on light squalls coming up. Looks hard for tonight.

Monday February 26th — Weather foggy. Under double reefed topsails. This eve put on whole topsails. Making E NE & wind hauling.

Tuesday February 27th — Weather fair same and occasional light fog. Carrying all sail & making E NE with a 3 knott breeze.

Wednesday February 28th — Weather fair & clear. Since last night 12 o'clock we have had the wind nearly aft. Steering NE by E & gradually freshening. Now we have a cracking breeze occasionally hauling a little to make us shift a stud sail or two. By the way we had a little mist this PM. Lest our fair wind might turn to a gale we have taken in F & Mizzen T G sails & sent down the Royal yard.

Thursday February 29th — Weather so so. Wind all sorts. As we feared, the wind freshened to a gale & at daylight we were lying to with nothing out but close reefed M T sail, M Spencer & F T Staysail. Furled the foresail for the first time since port & I think the wind blew harder than it has before on this voyage. While we scud it was impossible to sleep in the forecastle. More because grindstones, barrels, scrub brooms & other light articles were dancing a jig over head & the chests, pots & pans between. Hove for her rolling scuppers too. Commenced making sail at 11 AM & now have out M. T.G. sail. The wind hauls as it dies so that we are but little free on L tacks steering E NE. Although turned up we slightly stove the starboard boat.

Friday March 1st –- Weather fair. Wind light. All sail on wind. Knocked off this AM to SE. Tacked at noon & have come up to NE on S tacks. Bent another foresail & M Top sail. Repaired the foresail.

Saturday March 2nd –- Weather fair. Wind fresh for old sail. Took in the F & Mizzen T G sail & fly jib. Repairing sails. Wind dead ahead to run E tacked.

Sunday March 3rd –- Weather foggy. Wind fresh. All sail on S tacks. Wind has hauled so that we make N NE on S tacks. Caught a porpoise this PM.

**Monday March 3rd –- Weather fair. Wind light. Making E on S tacks. Tarring rigging.

** NOTE -- Apparently another mistake in the dating.

Tuesday March 4th –- All as yesterday.

Wednesday March 5th - All as yesterday save a little more wind.

Thursday March 6th –- Weather mixed with fog & rain. Running S on the wind under topsails & courses. Preparing oil lashings for NW.

Friday March 7th –- Weather wet with fog. Last night our head wind died whilst in perfect health. But we soon took another from the SW which has gradually hauled around to N NE, its strength increasing with its age. It is so strong this eve that we have taken off the T G Sng sail & Mizzen T G sails yet run 10 knotts in a S SE course. Lat. 30º 36' N Lon. 153º W.

Saturday March 8th –- Weather cloudy with squalls & wetter. Spouts all round. Much rain last night. Wind has hauled to E SE so that we lose our course.

We had a bit of a fight this eve between a Yankee & Portuguese. The Captain coming into the fray laid all the blame upon the Yankee (although it belonged to the other party) threatened flogging if the like occurred again & stopped his watch below for a week.

Heretofore all our difficulties have been settled with a slight skirmish & immediately forgotten, but now that the Captain has come in, all hands are dissatisfied & we must be rode for the rest of the voyage, which God grant be a short one.

Sunday March 9th –- Weather same. Wind increased until we have double reefed Topsails. Making S. Now that the Captain has come to his senses he has stopped the "Portuguese" watch. This satisfies all hands.

Monday March 10th –- Weather rough. Light squalls. Making S by E under double reefed topsails.

Tuesday March 11th –- Last night what wind we had, came altogether in gusts. Today it has been steady & moderating so that we have carried M T G sail this PM. Steering S SW. Expect to make land tomorrow if there is any afloat. Fixing "fixings" for port. The Captain, as is natural for him, has permitted the two combatants to come below again before their week is up.

Wednesday March 12th –- This morn we raised the land about 30 miles distant at sunrise. With a fresh breeze aft we ran for the passage to Maui. Got our ground tackle ready. Royal & Studding sails aloft but to no purpose. We could not raise wind enough to carry us in today so we are standing out to sea again to have another trial tomorrow. Another sail in like situation with us.

Thursday March 13th –- Weather fine. Wind light. Stood off so far last night that we have worked all day to get where we were last night & are now standing off again. The sail that we saw yesterday has gone in today & another has appeared. Clearing ships side. Got the Anchors off the bows this PM & lashed them fast.

Friday March 14th –- Weather fair. Wind light until towards eve when it considerably freshened. Stood off so far last night we were at a loss to find ourselves. The ship we saw last night showed us the way. We are now hard & fast off Lakaina with everything snug at 10 PM. We can see some sail to assure us that we are not alone. Cleaning & painting chains.

Saturday March 15th –- Weather fair but blowing fresh. Our watch had liberty ashore after 10 o'clock. The place is still the same as when we last left. Saw a hog baked whole (for description inquire of writer). The drunken sailors & Kanakas had a general fight. Ammunition was ground apples & shelalies. One sail went out & two came in which makes 28 sail. The King came in last night in a schooner from Oahu. This morn he went ashore with a salute from the fort of 21 guns.

Sunday March 16th –- Weather fine. Watch ashore. Watch on board gaming with other ships. One sail has gone out and two have come in. I received a letter from home last night dated May 23rd 1843.

Monday March 17th –- Weather fine. Our watch ashore. I remained on board to write a letter to Fish which is finished. Two sail came in and one has gone out.

Tuesday March 18th –- Weather fine. Larboard watch ashore. Painting ship & getting off wood. Five sail arrived, one a merchant brig the *Maryland* of Boston. One of the larboard watch came off pretty well educated.

Wednesday March 19th –- Weather fine. Our watch ashore taking in recruits. I had the good fortune to find a former roommate of mine Linten L. Torbert. Left second mate on the *Martha* & is now six months on the island working at his trade. He has just commenced making (?).-–— --- --- with a partner. Five sail out & one in.

Thursday March 20th –- Weather same. Larboard watch ashore. Getting off wood & water. Watch came off as before. Four sail came in & two have gone out.

Friday March 21st –- Weather cloudy & windy. All hands on board stowing water, getting off wood & taking bread ashore to be stored until our return. One sail departed.

Saturday March 22nd –- Weather fair. All hands getting off wood & water & coopering bread.

Sunday March 23rd –- Weather fine. Third watch ashore. I spent the day with Torbert & at church. Three sail departed & one arrived.

Monday March 24th –- Weather fine. Larboard watch ashore. Our watch stowing down water & as is usual can do more work & do it better & easier than with all hands aboard. It is now 11 o'clock & I have finished my day's work & that I may not forget what a day's work in port is, I will record --. At 4 o'clock just before day breaks all hands are called. Although sleepy & in the dark we find our way on deck in a few minutes. This won't do. A day's record will take all night so I must wait

The Log Book

for leisure & do it from memory. Watch came off pretty well drunk. Three sail arrived & three departed.

Tuesday March 25th –- Weather fair. All hands aboard staying forward the foremast & getting off our last recruits. Three sail arrived & three departed.

Wednesday March 26th –- Weather same. Getting off our last water & lashing it on deck besides getting everything else in readiness for sea tomorrow. Shipped a new cook as black as the ace of spades.

** Thursday March 28th –- Weather fair. Hove short at daylight but being calm, the Captain went ashore after a Kanaka. He got his Kanaka but while there one of his boats crew deserted him. Being unable to catch him he sent the boat aboard with orders to not let a forecastle hand leave the ship. The boat steerers had to pull after him & they even were not allowed to step ashore. We all hope that the fellow will not be caught for he is a useless tool. This eve a mate has to pace the quarter deck & a boatsteerer on the bows. All of us forward are laughing at the useless fear the Captain shows.

** NOTE-- This corrects mistake in dating on March 3rd.

Friday March 29th –- Weather fair. Hove out our anchor this morn & towed into a breeze & are now standing off & on Oahu. Fired 3 guns as we came out. Found out today that a stranger has stowed himself away on board us to get clear of his ship in Maui. What will be his fate I know not. Double reefed topsails this eve.

Saturday March 30th –- Weather fair. Standing off & on Honolulu. Captain went ashore this morn with a picked crew. I was chosen but chose not to go for several reasons. Left the old cook here & shipped two Kanakas. The runaway was taken ashore this morn in irons & left with the consul. We are now bound off, everything right & tight for the N West. Sold 5 pipes of bread.

Sunday March 31st —- Weather fair for the trades. Making N NW under double reefed topsails on the wind with a good top gallant breeze. The Captain distributed the clothing he bought in port today. To my lot fell a pair of boots & a pair of wool pants.

Monday April 1st —- Weather mixed with squalls of rain. Making N on the wind under M topsail & reefed fore & Mizzen. Making preparations for taking oil.

Tuesday April 2nd —- Weather fairer. Squalls lighter. Making N by E on the wind. Work as yesterday. Raised a sail this morn off our W beam. She has kept off until she is right ahead of us. Probably wants to test our sailing but no go. The Captain is in no hurry & carries a very easy sail.

Wednesday April 3rd —- Weather fair. Wind light. All sail set. Course N NE. A little free sail still ahead of us luffing up & then standing off across our bows. Painting fife rail & bits.

Thursday April 4th —- Course N by E & sail out of sight. All else as yesterday.

Friday April 5th —- Weather, wind & course same. Nothing new but wearing & tacking for finbacks & bending Mizzen Top sail.

Saturday April 6th —- Weather dirty with frequent rain or fog. Wind has increased since morn until we are under double reefed topsails. While furling the M T G sail this AM, one of the larboard watch lost his hold & fell as far as the top where he caught onto the thrash line & succeeded in recovering himself. Such small matters are, with us, a mere matter for a joke. No matter how great may be a man's danger, if he is not seriously hurt, he is laughed at. We shake hands with death too often to fear him much.

Sunday April 7th —- Weather fair with a M T G breeze. Caught a porpoise this morn. Lowered at 10 AM for 3 right whales but could not get on. Lowered again at 3 PM. In five minutes the waist boat was fast & at sundown we had the largest whale we have taken fast by the flukes. Sail in & tackles up.

Monday April 8th —- Weather fine. Wind light. Finished cutting by 3 o'clock & have started the works. Whales all around. Lat. 34° 30' N Lon. 155° W.

Tuesday April 9th —- Weather & wind as yesterday. Standing on the wind under easy sail. Boiling & setting up pipes. Finbacks.

Wednesday April 10th —- All as yesterday save speaking the ship *Europa* out of Bremen. Boiling & taking in all sail to night for fear of wind.

Thursday April 11th - Weather tough. At daylight the wind & rain came so strong in squalls that we were obliged to cool down & commence lashing oil. Before noon we struck a hundred barrels into the blubber room & lashed the rest solid. Started the works again this PM but had to stop again in a short time. We are now laying too under close reefed M T sail, M Spencer & F. T. staysail. Wind blows hard. Lee rail rolling under & the bulwark planks starting. Sail to windward this AM.

Friday April 12th –- "Tis morn yet scarce can you Lurid Sun" &c expose to us our situation above decks. All small articles scattered in all directions. A cask of oil gone. The bulwarks forward of the gangway all stove out. The starboard boat stove, the boat over the stern gone, the M T sail started, the F T Staysail gone & the running rigging flying in all directions. It is in fact blowing a "screecher". Stowed all the loose things between decks. The weather I think has moderated a trifle from last night but things look blue yet. Wind from N NW.

Saturday April 13th –- Weather cloudy & moderating. Commenced trying our last blubber this PM. I think our whale will stow 200 barrels. Had we had good weather he would have made us 240 barrels

Sunday April 14th –- Weather moderate. Sea going down. Made double reefed topsails. Used up the day in coopering oil preparatory to stowing down. Sail & finbacks to windward. Commenced standing boats crews. Took in all sail but M. T. & F sail.

Monday April 15th –- Weather fair with fresh breeze. Set F & Mizzen topsails reefed to steady the ship. Stowed our oil between decks forward in pipes. They have fairly installed me as coopers mate so that during a fare of oil I have but little to do but cooper. Wore ship this eve & are making N on wind on larboard tack. Standing whole watches tonight for fear of another blow.

Tuesday April 16th –– Weather blue with fresh breeze. Repairing ship & setting up pipes. Making N by W on larboard tacks.

Wednesday April 17th –– Weather growing worse. Put a fresh boat on starboard cranes. This PM hove too under close reefed M T sail. Took in the waist & bow boats & turned up the starboard boat. Stove the bow taking it in. While taking in the boats, the F Spencer gaff fetched away & came down on deck. Heading W by N on larboard tacks.

Thursday April 18th –– Still lying to but have set the foresail.

Friday April 19th –– Heading N NW under close reefed topsails. That's all.

Saturday April 20th –– Weather rough. Steering NW a little free under close reefed topsails & courses & jib. Nothing done but wet hold.

Sunday April 21st –– Weather fair. Carrying all sail. Wind has hauled to SE so that we are dead before it. Caught another porpoise.

Monday April 22nd –– Weather foggy. Reduced to close reefed topsails & foresail. Wind S. Course NW.

Tuesday April 23rd –– Wind hauled to NW last night & brought quantities of rain. Made sail to M T G sail. Stowed away bone.

Wednesday April 24th –– Weather clear & cold. Fresh T G breeze. Raised a school of sperm whales early this morn. Ran for them & lowered. Starboard boat soon struck a whale that ran with her until the line parted. Larboard boat killed a large whale. Bow boat fastened but after beginning to spout blood it took their line. The waist boat has had the hardest luck. First, carried away one sprit. Second, slightly stove but nearly run down by a fast whale. Third, run into by our whale, fastened another & got our line cut. Fourth, fastened to an ugly whale who stove slightly the boat where I was sitting. Fifth, went onto him again & as we struck him he knocked our bottom in.

We cut line after we found the boat filling & tried to keep her free but she soon settled with us to our waists. As long as she kept right side up we could stop in her, but she soon rolled losing every-thing that would not float. We managed to keep above water & on her for several rolls, but as we grew stiff from severe cold, it became more difficult to maintain our position.

Mr. Sanford, with my assistance, lashed two oars athwart the boat. Notwithstanding this, the boat again rolled bottom side up, rolling one young man from Connecticut (our bowsman) into eternity. I alone saw him stiffened as he was making an effort to get his head above water, but his efforts soon ceased & he sank almost within reach of me, but I was too far gone to help anyone but myself.

We remained on the boats bottom for some time, so long that we found that we could last but little longer. For my own part, I was as cool & self-possessed as ever in my life. Once I had made up my mind to let go my hold & thus shuffle off this mortal coil, all hope of present aid having fled. But then the feelings of my parents should I be lost at sea & my Father's last words "Never give up to despair" determined me upon making one more effort for our salvation & this was to right her & lash more oars athwart her, but how was I to do it?

Mr. Sanford was the only one from whom I could expect any assistance. The boatsteerer was in the bows frantically praying for assistance & too late repenting of his sins & promising amendment if saved. The other two were powerful Kanakas but too far gone to help us, besides one had his leg inextricably jammed through the bottom of the boat. We thought we could free his leg after righting the boat & save him.

After several attempts we righted her but the turn that saved us proved fatal to poor Louis.

His leg could not be freed & him too I had to see drown within reach of me. Mr. Sanford & myself worked hard (for we knew it to be our last effort) & finally lashed the steering oar and sail with the sails bolt rope. This done, I had just strength enough left to get cut a waife & set it. Mr. Sanford raised another amidships. These saved us. After setting my waife I sat or rather sank down into the stern sheets there to await my fate. Not fearful for after a trial I believe I can face death with any man. Once while most in danger I wished for a better recommendation to the other world but I knew it was too late to get it then, therefore made no trial deathbed repentance I do not believe in. Mr. Sanford seemed too well prepared for death. I thought he might have done more for our safety.

The last that I have any distinct recollection of was whipping with my hands various parts of my body & even the loggerhead to keep a little blood in circulation. At 3 PM I awoke in my bunk as from a troubled dream. Although buried under clothing with hot bricks at my feet, I was still shivering. I was told that the bow boat picked us up, stripping themselves to dry clothe us, that crew pulled on board almost naked. We were hoisted in & soon kindly cared for. When I awoke I could not move a limb without help but once moved they soon became limber save the shoulders & hips where the rheumatism still holds her seat. A thumb & two fingers are deeply cut or cracked, I know not which. The crew with sorrowful countenances are cutting in the whale. The Captain & all hands by turns have been down to see me. Lat. 45° Lon. 142°

Thursday April 25th — Cut in & commenced boiling. Weather rough. I & my brother sufferer have remained in our bunks. As near as

I can learn we were in the water about two hours & when picked up were blue faced & frothing at the mouth.

Friday April 26th –- Weather fair. Wind light. Under reefed topsails on wind. Boiling, setting up shooks & preparing boats & lines. We are to have another old boat. I have been on deck unable to do much for the cramp & sore hands. All hands feel more for the lost than we that saw then go for we were so near gone & every sensibility so thoroughly deadened that we could feel for none but ourselves. Stowed some time since requested in case he was lost that I should send his papers to his friends & I have now this mournful duty to per-form.

Saturday April 27th –- Weather clear. Wind fresh. Finished trying & stowed down most of our oil. The whale made 75 barrels. Took in sail this eve & stand boats crews.

Sunday April 28th –- Lying to in a moderate gale. This PM the wind went down. Set the F & Mizzen top-sail to prevent her rolling. Now lashing M sail & shaking out reefs.

Monday April 29th –- Hazy. Wind fresh. Running NW free. All hands stowing oil.

Tuesday April 30th –- Made double reefed top Sails this morn but reduced her to short sail again this PM, it being too rough to lower. Put out our boat this morn & are repairing the stoven.

Wednesday May 1st –- All as yesterday save that we furled the topsails this AM. Course NW a little free.

Thursday May 2nd –- Weather rough & cold with frequent snow squalls. Made close roofed topsails this AM. Took them in again this AM. Steering NW. Wind SW. For several days the weather has been colder than at any time last season. It requires all our clothing to keep comfortable.

Friday May 3rd —- Weather still blowing. Furled, loosed & furled the foresail again since 12 last night. M topsail & Spencers are all we have out. The Barque *Russell* drifted close by us on the S tack (we are on the L) & hoisted a signal we did the same.

Saturday May 4th —- Weather worse this morn. Took in the M spencer. This PM it has moderated so that we have set the spencer & F sail.

Sunday May 5th —- Weather fair. Wind has moderated until it is almost calm. Spoke the Barque *Pantheon* out of New Bedford, 3 months out with 800 barrels of sperm oil, no right oil as yet. The Captain went on board for thick clothing.

Monday May 6th —- Calm. All hands repairing & shifting sails. We have shifted M F & M T sails, Jib, fly jib & F T S sails. Grampuses all round. Lat. 52° d Lon. 148° W.

Tuesday May 7th —- Weather cold. Wind freshening. It now bids fair for a blow. I saw a right whale this morn, going quick to windward. Spoke the ship *Orien* from Nantes, France 8 months out, 500 barrels. Two sail at a distance. Have just finished repairing our stoven boats and are again in trim for whaling. Ran E while we could with all sail. Wind on S quarter.

Wednesday May 8th —- Furled the F sail last night. Set it again this morn. For the rest, a moderate wind, thick fog & nothing to do.

Thursday May 9th —- Fog, raining, snowing & blowing. Wind S W.

Friday May 10th –- Tried to stand lofts & carry close reefed topsails but no go. Had to take in & call down. Now under close reefed M T sail. Severe snow squalls. Wind W SW.

Saturday May 11th –- Stood whole watches last night to prevent the ships getting overboard. Made sail this A M & the wind has moderated so that it is now quite mild. Humpbacks & finbacks.

Sunday May 12th –- Weather fair this morn. Made all sail. Lowered at 8 AM for right whales but they were going too fast. Lowered this PM for another in a rough sea but missed him also. As soon as we came onboard we reduced sail to close reefed topsails & in a short time to close reefed M T sail. Blowing hard. Besides myself our boat has a crew three of the poorest men in the ship & consequently makes rough work at pulling. The Kanaka that was saved with me is still in his bunk from having his heel burned while recovering him.

Monday May 13th –- Lain to all day in thick fog. Blowing a strong breeze, almost a gale.

Tuesday May 14th –- Foggy & clear by turns. Wind has gone down. Made sail at noon. Raised whales to windward. Sail right ahead. Wind from SE.

Wednesday May 15th –- Calms & light winds. Whales to windward. Gamed with ship *Eagle* of Fair Haven 11 months out, 500 barrels. Just lost 3'd mate cutting a whale. Fog this AM. Drizzle this PM. Sail in sight.

Thursday May 15th –- Wet, foggy & blowing "on". Took in T sail this eve.

Friday May 17th — This AM foggy, wet & blowing hard. Fairing away this PM. Set F sail. Prospect of fair day tomorrow.

Saturday May 18th — Foggy & blowing. Captain has just cut us down from an indefinite quantity to 6 barrels of meat per month. A short allowance for this weather.

Sunday May 19th — Foggy, calm & comfortable. Made all but M sail but no lofts. The Captain, as becomes him, has today allowed us 9 barrels of meat if we behave ourselves.

Monday May 20th — Foggy & wet with occasional clearings & a stiff breeze. Carried reefed topsails, Jib & SP. Running Eastward. The Captain gave a man a cooling at masthead & all day on deck for sending him an insulting message. Humpback breach.

Tuesday May 21st — Tolerably fair. M.T.G. sail. Spoke with the ships *Gange* of Havre & *Orion* of Nantes. The former 5 months out and 250 barrels

Wednesday May 22nd — Weather good save for an occasional sprinkling. Wind light & astern. Steering E NE. Passed one merchant brig & 3 whalers.

Thursday May 23rd — Weather as yesterday. Steering N by W. Wind aft. Finbacks.

Friday May 24th — Weather pleasant & wind light. Since 4 PM we have taken a large cow and killed her calf. Larboard boat fastened. Nothing out of the common course save the whales being ugly while the calf lived & the cabin boy jumped overboard when the boat was on the whales. A cowardly Portuguese had a trick at bellowing every time our boat went onto the whale. We have made our fish double fast for it looks like a blow.

Saturday May 25th –- Commenced cutting after breakfast & finished by 4 PM. Boiling tonight. The Captain came near losing his life by falling overboard between the whale & ship. As it was he escaped with a badly bruised face. Spoke with the ship *Warren* of Do-- 10 mos. clean.

Sunday May 26th –- Foggy & rainy. Boiling. As a thing unusual we have had watch below while boiling.

Monday May 27th –- Weather mostly clear but occasionally a little rain. Set whole topsails this morn but took them in this PM. Blowing stiff in spots. Boiling. Finbacks.

Tuesday May 28th –- Weather decent. Wind fresh. Finished trying & stowed a part. The whale, although large, made but 75 barrels. Lowered but the whale got gallied by smoke. Ran for Finbacks.

Wednesday May 29th –- This morn turned out & furled fore sail & close reefed M T sail. It has since died away to a wet calm. Finbacks.

Thursday May 30th –- Weather fair & almost calm. All hands this AM stowing oil. This PM scraped bone. Finbacks.

Friday May 31st –- Weather fair. Wind light. Lowered 3 times to no effect. Finbacks in droves. Dried & stowed our bone.

Saturday June 1st — Weather as yesterday. Lowered once. The whales are all on the move. Carrying sail tonight. Working Northward. As it was my lot at Sunset I had an opportunity of witnessing a curious phenomenon. After the Sun's lower limb touched the horizon it assumed these several forms. (**at this point several pencil sketches have been drawn, however they were not with the manuscript.)

Sunday June 2nd –- Weather same. Lowered just at night for something that has gone somewhere. We are in Lat. 57º N. with 22 hours of daylight.

Monday June 3rd –- Weather same save wind lighter. Lowered at 4 AM & larboard boat fastened to a large & ugly whale. Bow boat got stove going on to him & we brought her aboard. At 9 we were just taking him alongside as the rascal sank. Lowered since for finbacks. Another boat on the bow cranes & the stove repaired.

Tuesday June 4th –- Weather same. Lowered at 4 AM. Bow boat fastened to a large whale. At 8 we had him alongside. At 5 PM we had cut him in and now are boiling.

Wednesday June 5th –- Some fog with light wind. Boiling.

Thursday June 6th –- Clear & calm. Finished boiling. Stowed some. Finbacks.

Friday June 7th –- Weather cloudy. Wind light. Commenced stowing this morn. Lowered twice this AM. Second time made an easy capture of a large whale to the bow boat. Commenced cutting after dinner & in 4 1/4 hours he was on board. Spoke skip *Warren* of Do??. Taken one whale & boiling it.

Saturday June 8th –- Weather cloudy with a little rain. Boiling & stowing. Made sail this PM & raised a whale going to windward. Sail in sight.

Sunday June 9th –- Weather cloudy & some fog with calm. Boiling. Lowered just at night for gallied whale. The *Warren* is about 2 miles. Just taken a whale. Finbacks, grampuses & porpoises.

Monday June 10th –- Weather cloudy. Wind fresh. Finished boiling this morn. Lowered twice this AM. The second time the larboard boat took another large whale. Began cutting at 2 PM & finished at 7 o'clock. While cutting, a large whale spouted several times within dart of the ship.

Tuesday June 11th –- Weather foggy. Made no sail. Boiling & stowing.

Wednesday June 12th –- Weather same. Wind fresh. No sail. Boiling & preparing casks. We can almost see the end of our voyage. Four whales more will fill us.

Thursday June 13th –- Weather fair. Wind light. Finished boiling this morn. Checked off the main hatch.

Friday June 14th –- Weather foggy. Wind light. Made no sail. Scraping bone. While we have been taking oil I have been constantly employed with the cooper. I have almost stolen the trade & have a great mind to steal the tools.

Saturday June 15th –- Weather foggy this AM. Wind light. Lowered just at night for a whale. Waist boat fastened but drew. Whales all round by the dozen. Finished bone scraping.

Sunday June 16th –- Weather rough. Wind fresh bringing occasional fog & squalls. Ran past a dead whale this morn & lowered for a live one too quick for us. Lowered again at 9 AM with same result. Lowered at 10 & after trying several, our boat finally fastened to an ugly 70 barrel whale.

He sounded once taking all our line and part of another & then commenced a race on top of the water. For 4 hours we darted lances & irons at his small without effect, the other two boats in vain endeavoring to fasten. Finally the bow boat fastened & we gave the larboard boat our line that we might take our Kanaka on board to save his life, for he was senseless & raving with suffering from wet & cold. We were obliged to hold him to prevent his jumping overboard. We went off again immediately & took our line. The larboard boat came alongside & took a fresh crew with an extra man & they soon put an end to the rascal. My bones tell me where I've been tonight.

Monday June 17th — Cut in by half past 9. Lowered twice, the second time the larboard boat captured a large whale. Commenced boiling at noon. Whales all round.

Tuesday June 18th — Weather good. Wind light. By 2 had the whale cut in and boiling. Lowered this PM.

Wednesday June 19th — Since morn it has been freshening with some fog & rain until it blows a small gale. This eve cooled down & lashed all solid.

Thursday June 20th - Weather fair in AM. Wet this PM. Stowed 80 barrels in fore hold. Boiling. Made sail at noon. Lowered once but not getting on easily, came on board. A ship to leeward just taken a whale. Plenty of whales all round.

Friday June 21st — Weather calm & warm. Lowered at 7 AM & larboard boat sunk a small whale. Lowered at 10 AM and the larboard boat fastened & drew. Lowered again for nothing and again this PM for something. The larboard boat took a 70 barrel whale & the starboard boat fastened to another but drew. Boiling. Sail in sight doing likewise.

Saturday June 22nd — Weather fair & wind light this AM. Foggy & blowing this PM. Larboard watch commenced cutting alone this morn. We turned out at 5 & finished cutting at 1/2 past 10. Made sail & ran for more whales. Lowered after dinner but as it was becoming foggy we came on board & took in sail & sent the watch below at 1/2 past 1 for their share of 15 1/2 hours sleep. Two sail in sight & whales all round.

Sunday June 23rd — Weather foggy & blowing. All sail in & boiling. We are so near full that the boat steerers have today moved into the forecastle & the cooper into the cabin. 200 barrels more will fill it. Spoke to the ship *Flora* of New London, 11 mos. out & 1400 barrels, 4 whales this season. Another sail in sight.

Monday June 24th — Rough & mostly thick. Stowing. Two sail, one cutting. Finished boiling.

Tuesday June 25th — Weather moderate & cloudy. Stowing. Lowered twice. The second time the bow boat fastened to a cunning chap that has finally left us with the line & irons of the bow and waist boats.

Wednesday June 26th — Weather fine. Wind light. Lowered at 5 AM & the starboard boat took another 100 barrel whale & got him alongside by 8 o'clock. Spoke to the ship *Orion*, 5 whales just cut in one. Three sail in sight.

Thursday June 27th — Weather wet. Wind light. Running S expecting to take one more whale & then off. Lowered this morn but gallied. Boiling & setting up shooks. Watch this PM.

Friday June 28th — Rough with some rain. Finished boiling this eve. Stowed 7 pipes. Watch below for every one save the cooper & myself.

Spoke to the ship *Hannibal* out of New London 9 mos. out & 400 barrels. Left their Captain in Oahu sick. The mate commands.

Saturday June 29th — Weather rough & wet. Stowed oil this AM. This PM lowered once. Fifty barrels more will fill us.

Sunday June 30th — Wind light & a little fog. We commenced whaling at 4 AM & finished at 10 PM. Have lowered 5 times for at least 50 whales. The 3rd time the larboard boat fastened to & soon sunk a large whale. The last time the larboard boat took the largest whale we have taken this voyage. Where we are to put all his oil I know not.

Monday July 1st — Wind light. Rain heavy. Cut in all but throat & lips & started the works. Sail in sight.

Tuesday July 2nd — Weather good save for little fog. Set up the last of our casks. Tomorrow we must pick up the pieces for all that will hold oil must be filled. Spoke to the ship *Sharon* out of Fair Haven 37 mos. out & 1500 barrels. Boiling & whaling. The *Sharon* lost her master a year since, killed by his Kanakas. Steering S & E on the wind. All sail set. Lat. 55° 37' Lon. 147° W.

Wednesday July 3rd — Weather wet with a stiff breeze from southward. Cooled down this PM to prepare more casks. Spoke to the ship ??? out of Greenport 1500 barrels. Boiling. Set the Main sail for the first time this month.

Thursday July 4th — Weather foggy but comfortable. Finished boiling & stowed all the ship will hold below deck. Every cask is full & we have to head up tubs to hold our oil. Have 50 barrels on deck. Wind light from the West. Course S SE.

Friday July 5th — Weather wet. Bent & set F T G sail this Morn. Wind S. All sail set on S Tack. Wind continued freshening & we were taking in sail until 6 PM when we reduced her to a double M & close reefed F Man T sails. Took in the waist & bow boats that we may carry the harder. The wind has hauled to W SW & our course is S by East.

M Spencer gaff came down this eve. Triced it up again & took in the sail. Scraped bone until there was danger of it washing overboard.

Saturday July 6th — Weather same. Wind little lighter. Made single reefed topsails. M sail, Jib & Spanker.

Sunday July 7th — Weather same. Wind still W. Steering S. Set M T G sail this morn. Soon took it in again. This eve reefed F & Mizzen topsail. Making about 6 knots. Lee rail is under & no safety save on the quarter deck. Now that we are once full, we can calculate with some degree of certainty upon the length of the voyage & all hands are laying plans of future operations. As for mine, so much depends upon circumstances that I can say nothing for the future save that my parents shall as soon as possible be satisfied of my existence.

Monday July 8th — Weather better. Wind lighter. Set M T G sail this PM & scraped bone.

Tuesday July 9th — Wind W. Course S by E. All standing sail set. Growing milder. Finished scraping & washed the bone. Put out the boats.

Wednesday July 10th — Weather same, wet. Wind has hauled round S SW fresh. Took in Fly jib this eve. Washing ship.

The Log Book

Thursday July 11th — Weather wet, wind high. It has been veering all day until it has arrived N & there it remains. Steering S with square yards. Nearly finished washing the ship. Set F T M S sail this eve.

Friday July 12th — Weather foggy & wet. Wind light. After boxing the compass it has finally become stationary at S W. Making S SE. Sent up M Royal this morn & made preparations for spreading all our kites to the first fair breeze.

Saturday July 13th — Weather fine. Wind light. Making S SW. Curing & stowing bone. Had a boat steerer aloft part of the day. The Captain is bound to have a full ship. We at present have one empty cask.

Sunday July 14th — Weather fair. Wind fresh. Making S SW. Stowed a little bone this morn.

Monday July 15th — Weather fair. Wind light this morn but freshening & hauling aft. Sent up & set all three Royals F & M T. G S sails. Shifted F & M sails putting on our best. The *Nimrod* is going now her best with the wind on the quarter. Every rag to it. Drying & stewing bone. Replacing ratlines etc.

Tuesday July 16th — Weather fair save some light squalls. Wind strong as we can carry all sail. Course S SW. Wind NE. Running new rigging.

Wednesday July 17th — Weather same. Wind lighter & hauling Eastward. Polishing ironwork etc.

Thursday July 18th — Weather very fair. Wind E & light. Course S by E ½ E. Polishing iron. Finished stowing bone.

Friday July 19th — All as yesterday save for scraping & varnishing stantions. As near as I can judge the Captain is trying by hard pushing some of his men to drive them out of the ship. However he says nothing to me so it is all the same.

Saturday July 20th –- Weather fair. Wind fair & freshening from the N NE. Course S by W ½ W. Lat. 27° N.

Sunday July 21st –- Wind strong from NE with occasional light squalls of wind & rain. Course S by W ½ W.

Monday July 22nd –- Weather fairer that yesterday. Wind Same. Course same. Scraping craft & fixing small affairs. Expect to make land tomorrow.

Tuesday July 23rd –- Weather same. We made land this morn. We ran for it all day and now are laying aback for daylight. Bent the cables.

Wednesday July 24th –- Weather fair & twixt trades, light winds & calms. We are now (Sundown) at anchor before Lahaina. Find two North Westers here full & one small whaler with 1100 barrels.

Thursday July 25th –- Weather good & hot. All hands washing, scraping, painting & varnishing.

Friday July 26th –- Weather fine. Little doing except getting off our bread & taking ashore all kinds of goods to sell. Had most of the PM to myself with Torbert.

Saturday July 27th –- Getting off potatoes & pumpkins etc. Intended sailing for Oahu today but could not get ready. We heave up tomorrow morn.

Sunday July 28th –- Weather fair. Got under way this morn at 7 o'clock & at 4:30 PM we were off Honolulu. We anchor outside the

The Log Book

harbor until wind & weather proves favorable. We find the *Magnolia* here with 4300 barrels She sailed a week before us from New Bedford. Brought 4 passengers down with us from Maui.

Monday July 29th — Wind being unfavorable this morn. We are still outside. Got some shooks from the *Magnolia* & have set up 18 short casks.

Tuesday July 30th — Wind fair. This morn the *Magnolia* hove up her anchor. Our boats, the *Warren*'s cutter & two merchantman's boats towed her in until a line could be run to about fifty Kanakas who were in readiness to haul the ship up to the harbor. As soon as she was in, we came off & commenced heaving up our anchor. We just got the anchor aboard when the trades came off & we paid out our chain again. This has been a great day with the Kanakas. The King came down from Maui in company with his officers & chiefs in four schooners to celebrate the recovery of their flag a year since. When the King's craft came in sight of the town, a salute of 21 guns was fired from a sloop of war & one from the fort in town.

Wednesday July 31st — The wind let & the boats helped to tow us into harbor. We are moored stem & stern within a stone's throw of the ship *Warren* & a few rods from the wharfs. Slopped up the sails & commenced painting spars. A great time ashore. Guns firing, flags flying, Sojers marching & cheering their King on to ???

Thursday August 1st — Starboard watch ashore. Larboard watch painting ship & Kanakas continuing to celebrate. As soon as I set foot on land I laid my course for Mr. Damon's house but found no anchorage there. Both himself & his wife have gone to Owyhee on a visit. Their boy, I learn is dead. I then went to his reading room & there spent most of the day hearing news from home. While cruising I had a fair opportunity of seeing the King, Queen, Queen's Mother, all the ship's officers & soldiery of the land. Saw them at their great dinner etc. Time will not permit a description of all the minutia of the celebration. Suffice it to say that in all things.

Friday August 2nd — Larboard watch off. Painting still. The *Warren* has been firing at a target. We can distinctly hear every ball whiz past our quarter.

Saturday August 3rd — Ashore again. I had not been there an hour before I found an old acquaintance, E. Whittlesey, just from home with letters from friends for me. Fired a horn for the first time since I sailed.

Sunday August 4th — Larboard watch ashore. Let go L Bower this eve to prevent us dragging ashore.

Monday August 5th — All hands painting, bending sails, coopering & filling water etc. Saw Whittlesey today shipping his goods for Maui. He leaves here tomorrow.

Tuesday August 6th — Our watch ashore. Saw Damon as he came ashore from Maui & Whittlesey as he went aboard for that place. Larboard watch stowing the rum & getting goods & spare trash ashore for the auction tomorrow.

Wednesday August 7th — Larboard watch ashore. Finished stowing Rum. Filled our water pipes etc.

Thursday August 8th — All hands taking oil ashore, getting off water & stowing 50 barrels of it down the main hatchway. Sent try works, pots & all ashore which finishes our whaling for this voyage. The *Warren* & *Magnolia* have sailed.

The Log Book

Friday August 9th –– Got off the remainder of our water & lashed it ready for sea. We have 6 casks of oil & about 30 of water on deck.

Saturday August 10th –– Our watch ashore. I dined with Mr. Damon & received from him letters & papers to all his friends in Salisbury. Brought off a lot of fruit & am now ready for sea. The barque *Vermont* of Mystic arrived outside. Left the NW to bring in two men hurt by a whale. A Brig left today.

Sunday August 11th –– Larboard watch ashore. Nothing more.

Monday August 12th –– Turned out at daylight to get under way. First hauled up the Kedge, then hove up one of the anchors & stowed it. Then ran a hawser to a ship in shore, hove up the other bower & warped her up to the other ship ready to make sail as soon as we took the pilot onboard. At 10 the Pilot come aboard. We fired two guns, made sail, let go our fasts & were bound for home with one passenger & two men sent home by the Consul. Our 3rd mate left us here to go as 2nd mate of the *Vermont*. The Trades have been so strong that we are obliged to reduce to double reefed topsails.

Tuesday August 13th –– Weather clear. Calm last night under the lee of Maui. Stiff top gallant breeze today till 2 PM. We are again becalmed under the lee of Owyhee. Course SE by E. Nearly make it close hauled.

Wednesday August 14th –– Calm all night. Light Top G breeze today. Making S SE.

Thursday August 15th –– Weather good. Stiff T G breeze. Making SE. At work on sails.

Friday August 16th –– Weather & wind same. Course & work same.

Saturday August 17th — Wind fresh. Took in T G sails & set again, trying to set up rigging but a little too much rain for comfort.

Sunday August 18th — Weather rainy. Wind SE. Wore ship this eve & stand Norward.

Monday August 19th — Weather squally. Wind fresh. Making SE by E.

Tuesday August 20th — Weather good. Wind fresh. Course same.

Wednesday August 21st — Almost calm. Setting up rigging. Preparing to break out & cooper tomorrow.

Thursday August 22nd — Rainy this morn. Fair since 7 AM. Turned up all hands at 9 AM to break out & cooper oil. I find it has not leaked much. Wind boxing all about, what little there is of it.

Friday August 23rd — All as Yesterday.

Saturday August 24th — Raining all the time. The wind is from all quarters, strong and light.

Sunday August 25th — Doldrums yet. Wind all sorts & in all directions. Occasional rain.

Monday August 26th — Rain morning & evening. Winds baffling. Finished coopering our oil in forward hold. Begin upon Main tomorrow.

Tuesday August 27th — Wind SE & fresh. Too fresh to cooper. At work on rigging.

The Log Book

Wednesday August 28th — Weather light. Coopered oil in M hatches. I think we have done for the voyage.

Thursday August 29th — Weather clear. Wind strong. S by E course, close hauled. Tore up all the sheathing save that amidships & scrubbed decks.

Friday August 30th — Weather same. Wind too strong for fly jib. Course S by E, 2 points free. Repairing Main Spencer. Crossed the line to day.

Saturday August 31st — Weather & wind same. Course do. Bent fly jib. Repairing sails. A low Island on our weather bow.

Sunday September 1st — Course S by 2 two points free with all the wind. T G sails will bear. Light squall this P M.

Monday September 2nd — Squally last night, today as yesterday in weather, wind & course. Got up T & M T G breast backstays.

Tuesday August 3rd - Weather fair. Wind fresh. Making S by E. Repairing sails.

Wednesday September 4th — Tacked ship last night & stood Norward for fear of an Island in this region somewhere. All else same.

Thursday September 5th — Weather, wind & course same. Repairing & bending sails & getting on our stoutest suit of canvass.

Friday September 6th — Weather good. Wind moderate. Making S SE. Repairing sails. Lat. 16° S Lon. 155° W.

Saturday September 7th –- Weather same. Wind hauled round so that we head S by E free. M Royal & F T S sails out.

Sunday September 8th –- Weather fair. Wind moderate & has hauled from NE to NW. Royals Top & Top G S sails out. Course S SE.

Monday September 9th –- Weather cloudy. Wind fresh from S SW. Making SE close. Repairing sails. Bit of a row between two boat steerers.

Tuesday September 10th –- Weather fair. Wind hauled this PM so that we set T Mast S sail. Steering S E by E. Sail mending.

Wednesday September 11th -- Weather fair. Wind W & fresh. Steering SE by E. Set lower stud sail this morn. Repairing sails & ropes. Land O.

Thursday September 12th –- Weather same. Wind hauling ahead. Took in S sails this eve. Just making SE close. Now working on rigging.

Friday September 13th –- Weather fair. Wind light from S SE. Steering by wind on S tacks. Renewing block strap & making sennit.

Saturday September 14th –- Weather good. Wind light & dead ahead.

Sunday September 15th –- Almost Calm. Head wind.

Monday September 16th –- Weather good. Wind fresh from E. Making from S to S SE.

Tuesday September 17th –- Weather cloudy. A little rain. Wind strong from NE. Steering SE by E course. This PM wind hauled to N. Then set the Royals & Studding sails. We're flying now!

Wednesday September 18th –- Weather rainy. Wind varying from N to NE. Making 8 & 10 knotts. Stud sails & Royals out. Course SE by E. I have for the week past been gradually gaining strength & appetite without knowing the cause. But today I have out with the "Yaller Janders". How to cure myself I know not, unless whaleman's remedies will do it via purging & puking.

Thursday September 19th –- Weather rainy. Wind strong from NE. Took in Royals & M T G Stud sail this PM. Sent down F & Mizzen Royal yards. This eve squared the yards. Wind N. Course same.

Friday September 20th –- Rain. Wind variable & lessening. Sent up Royal yards again. Almost calm this eve. Course same.

Saturday September 21st –- Rainy still. Wind light from N. Took in bow & waist boats & stowed them on the bearers. Preparing things for rough weather. Course same.

Sunday September 22nd –- Weather mostly clear. Wind NW & since morn has continued to freshen until we have just now double reefed Topsails & furled M sail. Carried stud sail until the T.S. beam went short in the iron. Sent down Royal yard & turned up quarter boats. Course E by S. Lat. 34° S Lon. 130° W.

Monday September 23rd –- Weather clear & squally by turns. Carry whole topsails & M T G sail. Course S by dead before the wind.

Tuesday September 24th –- Weather as yesterday. Carried Stud sails until almost night when we had to reduce her to double reefed topsails. The wind hauling to SW. Course same. Lat. 35° S Lon 124° W.

Wednesday September 25th –- Weather as yesterday. All sail again. Wind from S to SW & fresh. Course same. Am still below. The jaundice has left but there remains a difficulty of digestion & want of strength.

Thursday September 26th –- Weather same. Wind strong from N to NW. Course same.

Friday September 27th –- Weather same. Wind fresh from W. Stud sails on both sides part of the time. F & Mizzen Royals aloft & set again.

Saturday September 28th –- Weather same. Wind W and all we can carry, M T G sail too. Sent down R yards again.

Sunday September 29th –- Weather, wind & course same. Reefed topsails last night. Stud sails & Royal sails this AM. M T G Sails this eve. Lon 109° 21' W Lat. 10° 30' S.

Monday September 30th –- Weather foul last night. Wind from W with hail & rain. The first watch had to reduce her to close reefed M T Sail & F Gallant S sail. Scud under this till morn when it moderated so that made some sail this eve. Have M T G sail on her. This heavy weather has started the oil between decks & we have some difficulty in finding & chocking it again. Course S SE.

Tuesday October 1st –- Weather clear. All sail. Wind W. Breaking out to & chocking oil & fixing things for a gale.

Wednesday October 2nd –- Since last night the wind has been freshening from the S until we had to heave to this PM under close reefed M T sail. Furled the Topsail at sundown & for the first time at sea we have nothing out but M sails & T sail & enough at that.

Thursday October 3rd –- Weather as yesterday. Wind screaming from South & SW.

Friday October 4th –- Moderated a little. Made close reefed F & M T Sails & Fore sail. Making E. This eve took in F S sail again.

Saturday October 5th –- Weather better & wind lighter. Running whole topsails. Took in fly jib boom that had been snapped during the past gale.

Saturday October 6th –- Weather lowering. Stiff top gallant breeze from SE by S. Making E NE.

Monday October 7th –- Weather fair. Wind fresh from SE. Tacked ship this morn & again this PM. All sail out & all hands chocking oil.

Tuesday October 8th –- Weather fine & calm. This eve a breeze is springing up from the Westward. Course SE by E. Bent jib & fore sail & sent up Fore Spencer Gaff.

Wednesday October 9th –- Weather misty. Wind fair & fresh till morn from N W, before noon it hauled to S SW obliging us to take in Stud sail & Top Gallant sails. After dinner double reefed F & Mizzen topsails. Now have M T G sails on her. Making SE by E close. Stood my watch last night for the first time since I hove out three weeks since. I am still somewhat weak.

Thursday October 10th –- Weather cold. Occasional fog. Wind dead ahead. Tacked ship twice now on S tack. Making E at present. Repairing sails although it is so cold that one must whip his hands.

Friday October 11th — Weather good. Wind fresh & has hauled to W SW. Stud sails out. Bent an F spencer for the first time this voyage. Course SE by S.

Saturday October 12th — Commenced taking in sail at midnight last night & by 7 AM had reduced her to double reefed topsails. This PM took a reef in F sail & F topsail. Wind hauled from NE to W SW. Lat. 50° S Lon. 87°.

Sunday October 13th — Weather fair. Wind has fallen off to a light T G breeze & hauled to S by W. Course S E by E. Lat. 53° 9' Lon. 86°.

Monday October 14th — Weather clear. Wind strong from NW. Course E by S, true course SE by E. Stud sails on both sides. Lat. 55° S Lon. 84° W.

Tuesday October 15th — Weather foggy & clear. Wind strong from W NW. True course E by S. Set two T G Stud Sails this morn. Both in now. Lat. 56° S Lon. 84° W.

Wednesday October 16th — Weather foggy & uncomfortable. Wind N & strong. Commenced taking in sail at 9 last eve & by 3 this AM had her reduced to double reefed topsails & fore sails. After 7 AM made sail to topmast Stud sail. We are now double reefing again. Lat. 58° S Lon. 84° W.

Thursday October 17th — Weather cold with frequent snow squalls. Took in M sail & double reefed topsails last night before dark but before breakfast had topmast stud sail on her again. Wind fresh from NW. This eve the wind has hauled to SW. Course by compass NE by N. True course NE by E. We are now off the cape & consider ourselves as going home. Saw a sail bound westward.

The Log Book

Friday October 18th —- Weather cold with squalls of snow & rain. Reefed top sails last night. Today carried M T G sail over reefed topsails. Wind hauling from W to NW. M T G sail in. Course NE by N.

Saturday October 19th —- Weather rainy & uncomfortable. Last night close reefed F & Mizzen topsails & double main. Took in M sail and jib. This morn made all sail. Wind NW. This PM the wind has hauled to SW. Stud sails & Royal set. Course N NE.

Sunday October 20th —- Weather clear this AM. Rainy this PM. This morn the wind hauled to N NW & obliged us to take in F & Mizzen Top Gallant sails. This PM it fell off & hauled to W NW. Set T Stud sail. Course N NE.

Monday October 21st —- Weather fair. Wind strong from SW. Stud sails & Royals out. Course N by E. Caught 3 land birds this morn. Name unknown. Larger than a pigeon, smaller than a partridge & whiter than snow. The Captain claims two & all the rest are mine etc. Lat. 52° 54' Lon. 54° 48'.

Tuesday October 22nd —- Weather clear. Wind fresh from NW. Knocks us a point from our course. Hauling aft again this eve. Lat. 50º 12'.

Wednesday October 23rd —- Weather fine. Wind aft & has at last died away. Stud sails on both sides. Although there are pumpkins & potatoes left yet, they are stopped to us. Course N & N by E.

Thursday October 24th —- Weather fair. From last night's calm the wind sprang up from the North & has gradually freshened until we now have to reef topsails. Standing on larboard tacks.

Friday October 25th –- Weather foggy. Wind moderate from W to NW. Course N NE. Made Royal & top-mast stud sail this morn. Took in S sail this PM. Had a severe shower of rain & hail attended with thunder & lightening.

Saturday October 26th –- Wet & foggy. Wind since morn moderate. Just ahead our course N NE. Last night double reefed F & Mizzen Topsails. Made all sail again this morn. Whenever the weather will permit, all hands are at work with saw, files etc. working curious wood, Whales teeth & bone into boxes, canes, whips & boards & some of them are very well done. Lat. 44° 20'.

Sunday October 27th –- Last night calm & wet. Today weather very pleasant with a fresh breeze from SW. Stud sails & Royal to it.

Monday October 28th –- Weather same. Wind hauled to SE. & freshening. As yet all sail set. Three Royals aloft. Carried Stud sails on both sides part of the time. Repairing fly jib boon. Course N NE.

Tuesday October 29th –- Weather "decent". Wind SE & now all we can carry are M.T.G. sail. Course N NE. Lat. 33° 49'.

Wednesday October 30th –- Weather as yesterday. Wind same. Lat. 35° 56' Lon. 35° 30'. Course N NE.

Thursday October 31st –- Weather fine. Wind light from SE. All sail. Sent out fly jib boom. Course N NE.

Friday November 1st –- Weather fine. Wind light & dead ahead. Tacked ship this eve & are making NW. The weather has become sufficiently warm to work upon rigging & all hands are at it & probably will be until we arrive in cold weather again. Now that we draw towards home the Captain is growing better & better every day. Instead of two we are to have 3 duffs per week; also stewed apples as long as they last. Butter will come next. This oiling of old sores is a thing that is practiced in most ships with effect. Feed men high & treat them well the last month of a voyage & they will forget all they have suffered before. Had we had the provisions that were sold in port & that we get now as extra when we were working hard, 18 hours out of 24, they would have added much to our comfort & strength & stopped many a hungry, grumbling mouth; for then we were half fed, now we are over fed.

Saturday November 2nd –- Weather clear. Wind fresh from N NW. Stand on both tacks. Commenced wetting hold against the weather being comfortable warm.

Sunday November 3rd –- Weather this AM foggy. This PM raining. This AM wind light from N NE. This PM the wind has hauled round westerly until it is now SE. Topmast Stud sail out.

Monday November 4th –- Calm & foggy. Rigging. Scrub brooms.

Tuesday November 5th –- Weather fair & warm. Wind light from NE. Making N by W.

Wednesday November 6th –- Weather, wind etc. all as yesterday.

Thursday November 7th –- Weather cloudy. This eve rainy. Wind as before.

Friday November 8th –- Weather rainy. Wind baffling, varying from NE by N to E NE. This AM wore ship & are now on larboard tacks. Making SE by S.

Saturday November 9th –- Weather fair. Wind light from same quarter. Heading same.

Sunday November 10th –- Weather fair in AM. Rainy this eve. Wind light & varying from NE to NW. Course NE when we can head it.

Monday November 11th –- Rain, Rain, Rain & little wind (that little ahead) to console us. How long we are to knock about in this latitude is a problem that I should like soon to be solved or I shall grow tired of canvass as a locomotive power.

Tuesday November 12th –- Weather cloudy with a little rain. Wind light from N & NW on larboard tacks. Steering NE when we can.

Wednesday November 13th –- Weather fair. Wind moderate from N NE varying slightly.

Thursday November 14th –- Weather fair. Wind etc. the same.

Friday November 15th –- Do - Do - Do - Do - Tacked skip this PM.

Saturday November 16th –- Weather, wind etc. same. Making NW. We were called aft at 5 PM to witness a repetition of an event that occurred with us a year since, namely, the launching of the remains of a deceased seaman who died this AM at 9 o'clock. He was one of the men sent home by the Consul at Oahu. When he came on board he was well enough to work most of the time for himself but from the day that we sailed he gradually grew weaker & thinner until he died. What his disease was I know not but it appeared as though his organs of digestion had been destroyed by Mercury & the course of life he had pursued. Food did him no good when his stomach would take it & latterly he had taken none at all. A body so emaciated I never saw. Besides bone there was not 10 lbs. of him. Merely a breathing skeleton. While he lived he had the kindest care & attention. Someone with him all the time.

Sunday November 17th — Rainy. Wind in all directions until 4 PM when it came out from the southward. Lighted up enough this PM to show us 3 sail. One of which we spoke, vis; the *Nile* of New Bedford, 58 days out. Went on board & got papers that were a treat to us.

Monday November 18th — Weather fine. Wind SE fresh. Stud sails & Royals set. Course N NE. Two sail. Expect to make Trinidad tonight. Setting up rigging.

Tuesday November 19th — Weather rainy. Wind light from S to E. Hove to last night for Trinidad & have been in sight of it most of the day. It this eve bears SE from us.

Thursday November 21st — Rain. Rain. All rain. This AM wind light & knocking about in all directions. This PM came out from the S fresh. Course N by E. Last eve Captain Sherman came forward to find the watch & two happened to be below. One "Sojering" the other not wakened when the watch was called. He gave the first a rope's end with impunity but the other (a little knot of a feller) picked the skipper up & pitched him into the lee scuppers & there mounted him but let him up immediately & then the 4th mate separated them for which the Captain is down upon him as it leaves the impression that he was worsted by a boy.

Friday November 22nd — Weather fair with a stiff top Gallant breeze. Making N on the wind on an S tack.

Saturday November 23rd — Weather cloudy & a little rain. Wind fresh. Making N close. The Captain & mate had a growl this morn which ended in pulling the mate off duty. This noon he came on again with the understanding that he was to do as he saw fit.

Sunday November 24th — Weather squally. Stiff T G breeze. Took in F & Mizzen this PM. Making course N by E. Lat. 12° 23' S Lon. 31° 31' W. The last of our captured poultry left us today, the rest having given up the ghost.

Monday November 25th — Weather fair. Wind fresh from the E. Carrying F & T G Stud sails & 3 Royals. Ratting down. Ran alongside of a strange ship & exchanged Longitudes. An Englishman ran down to us but as soon as she saw our ensign she luffed to the wind again. Lon. 31° 22'.

Tuesday November 26th — Weather same. Wind same. Work same. Brig to windward. Lat. 7° 10' S.

Wednesday November 27th — Weather same. Wind & course same (NE by N 1/2). Ratting down & drying & bunching bone. Spoke the Brig *Olivebranch* of London. Carried away fly jib boom last night in a squall. Lon. 31° 6' W Lat. 5° 12' S.

Thursday November 28th — Weather fair. Wind moderate & E. Looked out for land last night. This morn kept away to the N. Rattling & tarring. Saw a bark this morn.

Friday November 29th — Weather same. Wind hauling S. Course N by W. Tarring rigging. Lat. 1o 24' S.

Saturday November 30th — Weather same. Wind light from S SW. Course N NW. Scraping & painting Iron. Saw a splendid meteor last eve.

Sunday December 1st — Weather same. Wind light & aft. Course same. Stud sails on both sides. Lat. 2° 10' N. Crossed the line yesterday for the last time this voyage.

Monday December 2nd --- Weather same. Wind & course same. Looks squally tonight. Repaired M sail. Pecked anchors & sent out fly jib boom. Lat. 3° 37'N.

Tuesday December 3rd --- Weather squally. Wind variable from SE to NE. Tonight light & aft. Course N NW.

Wednesday December 4th --- Weather fair through the day but squally at night. Wind fresh through the day E NE. Lat. 6° 11' N. All hands today shifting decks load. Decks comparatively clear having shooked 7 large casks.

Thursday December 5th --- Weather fair. Wind moderate from SE to E. Course N NW. Scraping & lashing up spare topmasts. Lat. 7° 40'.

Friday December 6th --- Weather fair. Wind light & baffling tending southerly. Nothing to do. We expect a gale as consequence.

Saturday December 7th --- Weather fine. Wind moderate from E NE. Studs & Royals. Attempted scraping waist but found it too wet. Lat. 10° 4' N.

Sunday December 8th --- Weather squally. Wind strong from NE. Exercise Royals occasionally. Course NW by N. Lon. 42º 42' W.

Monday December 9th --- Weather same. Wind, course, etc. same. Put another boat on the cranes. Lat. 15° 43' N Lon. 44°.

Tuesday December 10th --- Weather etc. same. Royals &, Top sails. Stud Sail. Scraping poles & painting a little. Lat. 15° 44' N Lon. 46° W.

Wednesday December 11th — Weather fair. Wind strong from E. Stud sails & Royals to it, making 8 knotts. NW 1/2 W. Painting spars, deadeyes etc. Lat. 17° 49' N Lon. 48°.

Thursday December 12th — Weather good, save a little squall. Wind strong from NE. Can just carry all sail. Painting a little & spinning a little & scrubbing decks a great deal. The Captain, Mate &c made a cart & harness for dogs & have them in training. Fine fun for boys. Lat. 17° 51' N.

Friday December 13th — Weather fair. Wind fresh from S. Course NW by N. Painting & cleaning small affairs. Lat. 21° 54'.

Saturday December 14th — Weather fine. Wind aft N by W course. Painting. Lat. 23° 50' Lon 55° 15' W.

Sunday December 15th — Weather fine. Wind light from S. Course N NW. Lat. 25° 21' Lon 56° 24' W.

Monday December 16th — Weather fine. Wind light from S but freshening. Still painting small pieces. Saw 3 sail. Spoke & got some papers from the *Star* (hermaphrodite brig) of & from New York bound to Barbados, 16 days out. Learned that the States have Porked themselves. Raised $35.00 by subscription for Enoch Mudge, Sea-Chaplain at New Bedford.

Tuesday December 17th — Weather fair. Wind strong S SW. Commenced taking in sail this eve. Reduced to top Gallant sails. Piping. Lat. 27° 51'.

Wednesday December 18th — Weather and wind are squally. Raised us out at 2 o'clock this morn to take in sail. Reduced her to fore sails & close mizzen topsail. The M T sail having split. This PM made M T

Sail & doused it again. Bent another M T sail & repaired rents. Since 9 last eve the wind has been NW.

Thursday December 19th –- Weather fine. Wind light from NW. Making SW on one tack & N on another. One sail & plenty of sea weed.

Friday December 20th –- Weather mixed. Royal at noon with courses. M T G sail now & making NW by N. Scouring little things. Lat. 31°12' N.

Saturday December 21st –- Weather clear through the day, tonight rain. Wind varying from SW to N NW. Fulls (?) bye. T G Sails. Made decks as clear as possible. Wet hold & chocked hatches for the last time. Lat. 32° 28' Lon. 62° 57'.

Sunday December 22nd –- Weather rainy. Wind strong from SE most of the time. This eve hauling westward. Course W by N while we could steer it. Now on the wind with M T G sails.

Monday December 23rd –- Last night double reefed topsails. This morn made T G sails. Now lying to heading NW under close reefed M T sail. While we ran our course was N NW. Saw a sail running N. Lat. 35° 0' N Lon. 66° 0' W.

Tuesday December 24th –- Still howling at 4 PM but since it has moderated & we have made F Sail & close reefed topsails.

Wednesday December 25th –- Weather fair. Wind light & has hauled to E. Topmast Studsail out. Saw sail last night. This morn picked up a topsail yard & sail that had been cut from some Brig. The sail had been split by the wind & it appeared as though they had been unable to take it in. Shifted M T Sail. This eve the Captain put a boat steerer in irons for not going to masthead for punishment.

Thursday December 26th — Weather is mostly good. Some light squalls. Light breeze since noon from S to W SW. Course N NW when we could head it. Washed ship outside this AM. Just spoke by moonlight to the Hermaphrodite brig *Napoleon* out of Portland. Three days out bound to Matanzas. Another sail in sight.

Friday December 27th — Weather fair. Wind fresh from SW by S. This PM hauled to S & freshening. Began to take in T G sails. Course NW by W. Have run out of gulf weed & some think we have crossed the gulf stream. Lat. 37° 0' Lon. 68° 19'.

Saturday December 28th — Last night the starboard watch had another opportunity of showing themselves at home in a gale, for at 7 o'clock she carried M T G sail & at 11 o'clock she was hove to under close reefed M T sails. Thus we still remain. Set the F sail today but furled it again. When the gale commenced, the wind was S but as we took in the F T sail it hauled to W NW & thus remained. The gale first came in squalls but it is now steady. Seen several pieces of lumber afloat. Probably part of some vessels deck load. Were it not that I consider what is termed "ill luck" my portion for life I might feel uneasy at being thus head off by the wind within one days sail of New Bedford & with the rest complain of our lot. But on the contrary I feel perfectly contented no matter what comes or how it comes. I take it like a philosopher determined that no cause which I cannot control shall disturb my temper. Last night all were sanguine in the belief that this was to be our last day at sea but now hope has almost disappeared in the distance for we may be wind bound for a month, perhaps more. For some time past I have been trying to analyze my feelings upon some subjects particularly that of returning home after so long an absence but I find it a difficult task. While we were on the other side of the world no one could be more anxious to return to a home still dear to the embraces of parents, brothers, sisters, relations & friends than myself, but since we doubled the Horn the desire to return has gradually given place to a fear that something has occurred which will occasion more pain than the greeting of friends can pleasure. Indeed I almost dread the idea of returning to Stamford where I may find myself without friends or home.

Sunday December 29th — Weather cloudy. Wind fresh from W NW to N NW. Wore ship this morn & working westward on the wind. Saw a barrel & several pieces of lumber afloat. This AM ran out of rough into smooth water wherefore can't tell. Weather since wind came out NW very cold. Need all our woolens.

Monday December 30th — Made all sail this morn & at noon commenced taking it in. Wind SW & constantly freshening. At 6 PM we had nothing out but close reefed M T Sail. Steering NW by W free. Much lightning this eve.

Tuesday December 31st — Wind hauled to NW by W. Making N by E under jib, M Sail & close reefed topsails etc. Wore ship this eve & are making SW. Sent down M Royal yard. Sail in sight. Standing with us. Lat. 38° 44' Lon. 63°. Saw a topmast & rigging.

Wednesday January 1st 1845 — Wind hauled to SW last night but this PM has hauled to the W again. Reefed F & mizzen topsail this PM but set them again with M T G Sail. Now they are double reefed again. Standing SW. We are now 150 miles from land and as bad off as if we were a thousand. Nothing but head wind gales. Three gales of 3 days each. The mate is fasting today & I suppose praying for a fair wind. A Hermaphrodite brig passed a bark to leeward. A ship to windward. Drift timber & specimens of ship carpentering frequent.

Thursday January 2nd –- Double reefed topsails all day. Making W by S on S tack. Wore ship this eve. We have run out our longitude & have now to wait a fair wind to make our northing.

Friday January 3rd –- Weather fine. Our N NW wind died away to a calm this AM & this PM a breeze sprang up from the Southward & still holds. Now making 7 knotts N 1/2 W. Hopes are again raised. I fear to be disappointed. Lat. 38° 41' Lon 71° 21'.

Saturday January 4th –- Wind last eve continued freshening until this morn when we were under double reefed T sails & fore sails. Making N. Sounded last night several times & found bottom in 35 fathoms. At 11 o'clock AM raised Montague Point & soon after Block Island. Kept away from latter under all sail. When abreast of the land a boat came off but had no pilot for New Bedford. This eve we lie aback inside of the Island under short sail with several lights in sight. Bent our cables etc.

END OF LOG.

Crew List

Captain

Wanton H. Sherman

Crew

Orrick Smalley
William P. Sanford
John Simmons
Samuel Minghan
----- Makee
Silvanus C. Tallman
----- Miller
----- Reynolds
----- Studley
William Giles
Andrew B. Beauvais
William Mason
Morgan H. Bradner
Graham Lee

Steelman B. Stowell
Charles Jacobs
David Connor
Thomas Peach
Joseph Thomas
William 0. Grady
Albert Tallman
Antoine -----
Emanuel -----
Richard Smith J
James Renn
G. S. Wright
Charles Wood
Justin Baker

Graham Lee's shipboard will.

For S. C. Tall-man

I wish you to fill my chest with my best clothes, books & all my small goods that are of any value and take charge of them until your arrival in New Bedford & then to write to my father explaining the circumstances. If he does not send for them the whole is yours. Destroy every letter in my chest and after you have taken what you want of all that is mine, give the remainder to the crew & by so doing gratify the feelings of my friends & greatly oblige one that was –

Graham Lee

Any one writing my friends will address:
Elisha Lee
Salisbury - Conn. U. S. A.

Bill of Sale

Monday 5th February 1844

Sold to Silvenus C. Tallman an Epitome scale & dividers to be paid for in kind or in cash to the amount of their value at home when new. The choice mine -----

APPENDIX A
GRAHAM LEE BIO

Graham Lee
brother to Henry Lee

Graham Lee, the subject of this sketch, was born January 22, 1821, at Salisbury, and was the son of Elisha and Almyra (Scoville) Lee. The Lees trace their ancestry to John Lee, born in 1621, a native of Colchester, England, who came to America in 1634, in the ship *Francis*, under the care of William Westwood at the age of 13 or 14. He located in

Cambridge, Massachusetts, where he remained a short time when he moved to Hartford, Connecticut, where he married and became one of the permanent men of the settlement.

Graham Lee was born 22 January 1821 on the same farm as was his father, and in the same house, where he was reared to the age of twelve, when his father moved to town and engaged in the mercantile business. Here Graham received a fair education and attended Lennox Academy, studying Greek and Latin in preparation for college, but his eyes were failing and he had to give up his studies.

At the age of nineteen he went to New York to superintend the dairy farm for his father where he remained till he arrived at the age of 21. On 15 November, 1842 he left on a whaling voyage on the whale ship *Nimrod*, out of New Bedford, around the Horn to the North Pacific Coast where they spent two years. On 5 January, 1845 the *Nimrod* arrived back in New Bedford and Graham took his earnings and went home. Later that year, along with his brother Henry, they walked to Ohio where they purchased 1,000 sheep and herded them to Mercer County, Illinois, to what is now Perryton Township. They laid a claim on section 9, which is now owned by him. He has made one of the most beautiful farms in Mercer County.

In 1853 he was married to Mary A. Candor, born in 1834, and a native of Union county, Pennsylvania. She came with her parents, Thomas and Margaret (Montgomery) Candor, to Mercer County, Illinois, in 1837. They had nine children. His wife, Mary died January 30, 1874.

He was married a second time, to Anna Sarah Fisher, a native of Greene county, Pennsylvania, born March 1, 1847. They had 5 children together.

Graham Lee died on February 11, 1908. He is buried in the Hamlet Cemetery.

APPENDIX B

Vessels and Terminology

An introduction to the essential tools – the vessels – and terminology used aboard a whaling voyage.

A Brief Look at Shipboard Terms

A whaleship was a floating community with its own rules and language. Landlubbers (land dwellers or new seamen) may need a guide to shipboard lingo:

Abaft: To the rear of or in the direction of the stern (rear) of the ship.

Aft: At, near, or toward the stern (rear) of a vessel; opposite of forward.

After House: The name given to a square or rectangular cabin built on deck near the middle of a whaleship. It was used as a place to get out of the weather or as a privy.

Aloft: Above the deck in the rigging.

Amidships: In the middle of the ship.

Boom: A sturdy pole, attached to the foot (bottom) of a fore-and-aft sail (see below), used for spreading and maneuvering the sail.

Bow: (Pronounced as in "take a bow"). The front end of a boat or ship.

Bower: The forward end of a vessel. Either side of this forward end, especially with reference to the direction of a distant object: a mooring two points off the port bow.

Braces: Ropes to move the yards in a horizontal plane.

Cat an anchor: To **cat** the **anchor**, to hoist the **anchor** to the cathead and pass the ring-stopper. To fish the **anchor**, to hoist the flukes to their resting place (called the bill-boards), and pass the shank painter.

Cathead: A cathead is a large wooden beam located on either side of the bow of a sailing ship, and angled forward at roughly 45 degrees. ... The stockless anchor made the cathead obsolete. In common practice, the projecting end of the beam was carved to resemble the face of a lion or cat.

Clew up: - clew up, Nautical. to haul (the lower corners of a square-rig sail)

Clew down: - clew down, Nautical. To secure (a sail) in an unfurled position.

Crow's Nest: Originally a barrel lashed at the top-gallant mast (the highest section of the mainmast) where a man was stationed to look for whales or ice. Usually only employed in the Arctic and Antarctic fisheries.

Deadeyes: a circular wooden block with a groove around the circumference to take a lanyard, used singly or in pairs to tighten a shroud.

Decks: The "floors" of a ship.

Fluke(s): Modern **anchors** for smaller vessels have metal **flukes** which hook on to rocks on the bottom or bury themselves in soft seabed as did the sailing ships of the 1800s. The vessel is attached to the **anchor** by the rode (commonly called cable when made of rope, and made of chain in larger vessels), or a combination of these.

Forward: Opposite of aft; front section of vessel. Fore: Indicates part of the hull, rigging, or equipment located at, near, or toward the forward end of a ship.

Fore-and-aft-rigged: A method of hanging sails on vertical masts at fore (forward) and aft (rear) so that they hang parallel with the keel of the ship (instead of hanging horizontally across the deck, as square-rigged sails do). Fore-and-aft-rigged ships were popular with owners because they required smaller crews than square-rigged ships.

Gallied: worried; flurried; frightened.

Gam: An exchange of visits at sea by the crews of two or more whaleships. The Gamming Chair was used to transport individuals from one ship to another.

Grampuses: A common name for the orca.

Kedge: - move by hauling in a hawser attached to a small anchor dropped at some distance.

Keel: A long structural timber running along the outside of the bottom of a ship from front to back - "from stem (another nautical term for front) to stern" (back or rear).

Leeward: Pronounced "loo' ard." The side away from the prevailing wind.

Lobscosuse: A type of lamb or beef stew. The word comes from "Lobscouse", a stew commonly eaten by sailors throughout Northern Europe, which became popular in seaports such as Liverpool.

Martingale guy: A stay running from the end of the jib-boom to the dolphin striker, which holds the jib-boom down against the pull of the fore topgallant mast stay.

Mast: An upright pole for supporting sails and ropes. A mast may be a single pole or number of poles in consecutive extension, one on top of the other. Each mast has a name determined by its height, such as "lower mast" or "topmast," or its position, such as the "main-mast," which was usually the second mast from the front of a three-masted ship.

Mollymock: The mollymawks are a group of medium-sized albatrosses that form the genus Thalassarche … The word mollymawk, which dates to the late 17th century, comes from the Dutch mallemok, which means mal – foolish and mok – gull.

Mother Carey's Chickens: Of nautical origin, the term Mother Carey's chicken designates the storm petrel. The storm petrel was thought by sailors to be a harbinger of bad weather sent by the Virgin Mary

Oakum: is a preparation of tarred fibre used to seal gaps. Its main traditional applications were in shipbuilding, for caulking or packing the

Vessels and Terminology

joints of timbers in wooden vessels and the deck planking of iron and steel ships; in plumbing, for sealing joints in cast iron pipe; and in log cabins for chinking.

Port: The left side of a ship, as the steersman stands facing forward. In earlier times, called "larboard."

Rig: The distinctive arrangement of masts, rigging, and sails that indicates a type of vessel, such as a bark or schooner.

Sennit: A type of straw which is braded to make ropes and bumpers for the ship.

Rove: To pass (a rope or the like) through a hole, ring, or the like.

Sennit: - Flat braided cordage that is used on ships. Cordage, the ropes in the rigging of a ship.

Shook: A set of parts for assembling a barrel or packing box. [Probably from shook cask, variant of shaken cask, cask broken down for shipment, from shaken, dismantled and packed for transport, past participle of shake, to scatter, shed.]

Slewer: Although the term "preventer" is more generally used in the Provincetown fleet, New Bedford record their extra harpooners as "preventer boat-steerers"; but the crew invariably call them "spare boat steerers." The terms "spare" and "preventer" are employed for anything held in reserve. The term "boat-steerer" owes its origin to the fact that the harpooner, after striking the whale, takes the steering oar and so directs the movements of the boat as to enable the officer to kill the whale. The term "slewer" a slang expression, is also sometimes used.

Spar: A general term for a strong pole used in the rig of a ship. Depending on its position and use, a spar may be called a boom, gaff, mast, yard, etc.

Spyglass: A small telescope often used by the captain on the bridge.

Square-rigged: A ship on which some of the principal sails are square in shape and hang across the deck, rather than running with the keel (as in a fore-and-aft-rigged ship.)

Starboard: The right side of a ship, as the steersman stands facing forward.

Stern: The rear of a ship.

Tack: To sail a zigzag course, as nearly as possible into the wind, to reach one's destination. (A ship cannot sail directly into the wind.)

Tonnage: The carrying capacity of a ship (not its weight).

Triced: To trice up is to draw up, shorten, or tighten some sail or rope.

Ware ship: is a sailing maneuver whereby a sailing vessel reaching downwind turns its stern through the wind, such that the wind direction changes from one side of the boat to the other. For square-rigged ships, this maneuver is called **wearing ship**.

Windward: The side against which the wind is blowing. Yards, horizontal poles which cross the mast and support the sails on a square-rigged vessel. The ends are known as "yardarms."

Sails

studding sail: A studding sail is an extra sail hoisted alongside a square-rigged sail on an extension of its yardarm. It is named by appending the word studding to the name of the working sail alongside which it is set (e.g. "fore topsail studdingsail"). These sails provide extra speed in fine weather.

top gallant sails: On a square rigged sailing vessel, a topgallant sail (topgallant alone pronounced "t'gallant", topgallant sail pronounced "t'garns'l") is the square-rigged sail or sails immediately above the topsail or topsails. It is also known as a gallant or garrant sail.

jib: a triangular staysail set forward of the forwardmost mast.

Royal Sail: A royal is a small sail flown immediately above the topgallant on square rigged sailing ships. It was originally called the "topgallant royal" and was used in light and favorable winds. Royal sails were normally found only on larger ships with masts tall enough to accommodate the extra canvas.

main top sail: A topsail set on the mainmast.

fore main sail: A foresail is one of a few different types of sail set on the foremost mast (foremast) of a sailing vessel: A fore and aft sail set on the foremast of a schooner or similar vessel. The lowest square-sail on the foremast of a full rigged ship or other vessel which is square-rigged.

Main Royal: A royal is a small sail flown immediately above the topgallant on square rigged sailing ships.

main top sail: A topsail set on the mainmast. Sail, canvass, canvas, sheet - a large piece of fabric (usually canvas fabric) by means of which wind is used to propel a sailing vessel.

mizzen top gallant sail: The square-rigged sail or sails immediately above the topsail or topsails. It is also known as a gallant or garrant sail.

double reefed: Reefing is the means of reducing the area of a sail, usually by folding or rolling one edge of the canvas in on itself.

Types of Whaling Ships

The Yankee whaler was a highly evolved vessel that incorporated a variety of technological details that served to distinguish it from any other type of craft. It was designed to carry a large crew of men (up to 35 individuals) who would process and store materials obtained in the hunt over a period of years. Here it must be said that not all whalers were built for the purpose of whaling. Many were converted to whaling from their previous uses in the merchant service. All whalers, regardless of previous use had various details making them unique. The most conspicuous feature was the brick furnace called the try works located just behind the foremast. Whalers also had three to five whaleboats hanging from big wooden davits on both sides of the vessel; two upside-down spare boats sitting atop a wooden frame mounted on the deck, and a deep and capacious hold where the large casks of oil could be stored. At sea a whaler could be distinguished by its slow speed, possibly a plume of smoke rising from the try-works and the men stationed at the top of each mast looking out for whales. While cutting-in a whale the large and heavy industrial-grade blocks and the men standing on boards over the side of the vessels wielding long-handled spades and the large group of men in the bow of the vessel heaving at the windlass marked the ship as a whaler. Many whalers were painted with false gun ports for purposes of disguise and intimidation from a distance. This painting scheme could deter pirates on the high seas or hostile peoples encountered at the many re-mote landfalls commonly frequented by these vessels. The average square-rigged whaleship was about 100 feet long and 300 tons carrying capacity.

Types of whaleship rigs:

Ship
This type of vessel has three masts, each with topmast and topgallant mast and square-rigged on all three masts. Ships often carried four boats, sometimes five and had the largest number of crew. There were six men per boat plus the ship-keepers, men who stayed aboard the vessel when the boats were down after whales. Ship-keepers included the steward, cook, cooper, blacksmith or carpenter. There could be as many as 37 people on board a ship.

Journal of a Whaling Voyage

This photo is of the "*Charles W. Morgan*" as she appears today, moored at Mystic Seaport, CT. This is the general ship type of the *Nimrod*.

Barks

Very similar to a ship rig in that it was a sailing vessel with three masts, square-rigged on the fore and main masts and fore-and-aft rigged on the mizzenmast. This rig became very popular in the mid-19th century as it required fewer crew to handle the sails when the boats were down for whales, thus saving the owners money.

Types of whaleship rigs:

Brigs
The true brig is a two-masted vessel square-rigged on both fore and main masts. Brigs were most often employed in shorter voyages to the Atlantic Ocean and saw use throughout the 19th century.

Hermaphrodite brig
A two-masted sailing ship which has square sails on the foremast combined with a schooner rig on the mainmast (triangular topsail over a gaff mainsail).

Schooners
The schooner was the smallest of the whalers, usually with two masts and four-and-aft rigged sails and carrying two or three whaleboats. Six months was the ordinary length of voyage and most schooners were employed in the Atlantic. Although the schooner was employed throughout the history of Yankee whaling it was especially favored in the later period (1890-1925), be-cause it was economical to outfit.

The Whaleboat

Classic design

Whaleboat builders refined the craft's design, shaping it into a dependable means of getting close enough to a whale for the kill. A whaleship sailed with three to five whaleboats swinging from davits (cranes used on ships). Spares, usually two, were stowed on top of the after house at midship.

Each whaleboat was:

- Light and strong
- Approximately 30 feet long, six feet wide
- Pointed at both ends
- Sometimes painted in bright colors at bow (front) and stern (rear) for easy identification at a distance
- Equipped with mast, sail, and rudder, as well as oars and paddles. The oars were unusually long, ranging from 16 to 22 feet long.

Sleek lines gave these boats beauty as well as speed and maneuverability. Their uncomplicated design made them easy to repair - important on long voyages, be-cause whaleboats were often damaged during encounters with whales.

Journal of a Whaling Voyage

Each whaleboat had a crew of six. The boatheader, usually the captain or one of the mates, stood on a narrow piece of wood across the stern (rear), handled the steering oar, and commanded the boat; The harpooner or boatsteerer pulled the bow oar up front and four crewmen rowed with oars that were balanced in length so the boat could be rowed equally well by four or five men.

The right equipment was essential.

Each whaleboat typically carried:

- Two wooden tubs, each with 150 fathoms (900 feet) of coiled hemp line. Care was taken to ensure that the rope would uncoil without kinks -- to prevent injury or death for crewmen or loss of the whaleboat.

- Two harpoons, ready for use, and two or three spares.

- Two or three lances, or barbless blades, used to kill the whale.

- Hatchet and knives to cut the line in an emergency.
- Wooden keg for drinking water

The Whaleboat

- A piggin – a small bucket for bailing water from the boat or for wetting the line attached to the harpoon in the whale, if it began to smoke when the line ran out rapidly at the wounded prey tried to escape.
- A lantern-keg with flint, steel, box of tinder, lantern, candles, bread, tobacco, pipes
- Compass.
- Waif - a long-poled flag used to locate a floating carcass from a distance and to identify it for other whaleships.
- A dragging float to make it harder for the whale to swim.

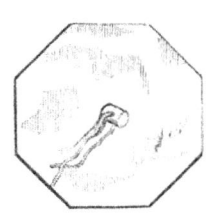

- Fluke spade to cut a hole in the whale's tail and tow the carcass back to the ship.
- Miscellaneous equipment, including anchor, buoy, etc.

APPENDIX C

The Varieties of Whales

The following descriptions are limited to the species that were most commonly hunted in the American whale-fishery:

Toothed Whales (suborder Odontoceti)

Sperm Whale (Physeter macrocephalus): Grows to up to 60 feet long, weighs up to 63 tons. Follows its food supply through the world's oceans - is generally found in colder seas in summer and in temperate and tropical waters in winter. Feeds on small fish, squid, giant squid. Dives to depths of at least 3,300 feet -- deeper than any other marine mammal. Holds its breath while submerged for up to 90 minutes. Displays enormous teeth on its lower jaw. Was the principal prey of the nineteenth century American whale fishery. Haunted Captain Ahab in the classic American novel, Moby-Dick.

Baleen Whales (suborder Mysticeti)

Baleen whales do not have teeth. Instead, they are distinguished by baleen, which hangs in strips from the roofs of their mouths. Baleen is composed of keratin, a substance found in nails, claws, horns, and hoofs. It looks like hairy, vertical venetian blinds. The whale uses it to strain out krill (masses of small shrimp-like crustacea that float near the water's surface) from sea water.

Right Whale (Northern Right: Eubalaena glacialis and Southern Right: Eubalaena australis): Grows to up to 60 feet, weighs up to 100 tons. Migrates through temperate waters from Florida to southern Canada.

Known as the "right" whale to hunt, (it was often close to the beach, visible to its land-based hunters and pro-vided a large supply of blubber) it moves slowly and floats after being killed. It was pursued first by both Europeans and Americans. It is the most endangered of all whales, with a total population probably not exceeding 400.

Bowhead Whale (Balaena mysticetus): Grows to about 60 feet in length and weighs 100 tons or more. Prized by whalemen for quantity and quality of its blubber and baleen. Carries the thickest blubber of any whale (20-28 inches), an adaptation to the icy Arctic waters in which the species lives. Possesses longest (10- to 14 feet) and largest number (600) of baleen plates.

Gray Whale (Eschrichtius robustus): Grows to up to 48 feet long, weighs from 25-30 tons. Migrates 12,000 miles roundtrip -- longest migration of any whale species - from the frigid waters of the Bering and Chukchi Seas, where it summers, to the warm lagoons of Baja California, where it winters. Considered ferocious by whalemen, who called it "devil fish." Present almost affectionate interaction between whale watching humans and gray

Rorqual Whales

The Varieties of Whales

Humpback Whale (Megaptera novaeangliae): Grows to up to 50 feet and weighs up to 50 tons. Does not have a hump but arches its back when it dives, which may account for the name. Displays huge flippers, which are nearly as long as one third of its body. Breaches dramatically, propelling its huge body almost entirely out of the water and diving back in with an enormous splash. Noted for complex, repetitive vocalizations. The Humpback was one of the five species normally hunted by the Yankee whalers, although it was the least desirable since it sank about half the time after being killed and its baleen was useless.

Blue Whale (Balaenoptera musculus): Grows to a length of 100 feet and weighs up to 150 tons, the biggest creature that ever lived. Because of intensive whaling in the 20th century, the Blue Whale has been left as one of the most endangered species. It was never hunted by the Yankee whaleman because it was considered too fast, too big, and because it invariably sank when killed.

Fin Whale (Balaenoptera physalus): Grows from 60 to 85 feet long and weighs up to 80 tons. Considered one of the fastest of marine mammals, swimming at estimated speeds of up to 25 miles per hour. Not hunted by whalers in the age of sail because harpoons became dislodged due to its swimming speed and, like its close relative, the blue whale, it usually sank when killed.

APPENDIX D

Letters written to Graham from family and friends during the voyage. In letters from M. H. Fish there are nicknames used at times i.e. Roderick for Graham and Furgeson for M. H. Fish. These letters speak, at times, of the political issues of the time as well as family and friends.

Letter from Myron Holly Fish to Graham Lee—M.H.S..., Ship *Nimrod*, Capt. Sherman, Care of Barton Rickerton, Esq., New Bedford, Mass.

In my den at Riga, May 8, 1843

My dear Roderick, (NOTE: a secret name, some romantic boyish notion)

Yours of Jan. 12th was taken from the office yesterday and if mortal was ever pleased at hearing from other I was one. You cannot imagine how long and anxiously I have waited for a letter from you and when at length it came it fully realized my expectations. You are so well suited with your business and upon considering the opportunities you have of seeing and learning places, objects, etc. that one does not meet on land, it is no wonder that you are so well suited. I have been sorry that I did not accept your proposal of starting for Astoria this spring and do not know that I have entirely lost the feeling as yet. My own affairs in regard to my situation to the Iron Co., etc., have not met the view I was led to entertain and consequently my situation is not a very pleasant one! Unless there is a very material change next spring, I shall leave Riga for some other place.

You no doubt expect that I shall give you a summary of everything new that has transpired in S. since you left. About two months ago I mailed a letter to you to N. Bedford and in that gave all the news up to that time.

May 20th: Since writing the above I have been busy in surveying and other business so have not had time until tonight to continue my letter. A day or two after I rec'd your letter I called upon your Father's and read such parts of it that I thought proper. They were very much pleased to hear from you, and as I observed, had become reconciled more to your absence. Still they all indulge the utmost anxiety that you will make Salisbury or someplace near your future residence. I have not the least doubt but if you had appreciated fully their feelings before you left you would not have sailed. They are all as well as usual. Fanny is teaching the district school in F. Village and Julia has a select one at the same place. Matters move along much after the old manner there and also here. In politics S. is yet low. H. Scoville and R. Averill, Esq. represent us in the Legislature this year.

May 28th, Sunday evening: After walking up the same old Mt. again, I will continue my letter & bring up all the news to this date. Within the last 18 days W.F. Bosworth has died, making the third death in the family within 3 months. Truly "in the midst of life we are in death." How strangely it falls upon the ear to hear of our friends and acquaintances

Appendix D

one after another falling off. But no less strange than true. May we all be prepared for our exit when the summons shall come. Since February the people of S. have been quite agitated regarding the "second advent" by the lectures of Mr. Crittenden from Hartford, and Enthusiast of the Miller School. His lectures produced quite an excitement wherever he lectured. But it has subsided in a measure.

Mr. H. Belcher has declared bankrupt in the sum of $2000 more than he is able to pay. It is said by many that it is a rascally affair, indeed it looks like one very much. The young people, God help them, are flourishing as usual. L. Bartell has been jilted by Miss Mary Whittlesey. A. Jasell has had a "responsibility" added to the cares of himself and his wife. They are both in poor health but the "Baby" doing well. The D. family are as usual. Both "take time by the forelock" and enjoy themselves as well as ever. He desired to be remembered to you. Coffing is as well as usual and sends his regards, also all the other young folks. Lee Hollister & Eaneas Lee stayed with me a few nights ago. E. is still studying at Ticknors, and L. is farming. J.B. Elliot has gone to Sharon. From your letter I should think that you don't have too high opinion of him. He is a good fellow though. But there is something lacking.

Next in order comes marriages. Miss Joselyn has married—Mr. Townsend, U. Trait's former husband. Miss Marie Wardwon, a Mr. Wallace of N.Y. And that is all I believe. As for our much respected friend the Knight, he has taken a voyage to Cape Cod, fishing for his health. But most likely, the true reason is to fill up his empty pockets. Some say that he because he thought himself too young to start practice of law. If so, and he continues to adhere to that very good rule, I think that he will not practice very soon.

Misses F & H are both in town, and the one grows more crooked sideways the other improves her looks by the assistance of art. I have never spoken since our last letters. But have received several intimations that all things were easy for a reconciliation. But the absence of one in the "Grave Yard" still haunts my memory, and although I have forgiven her the sin still recollects the insult.

June 3rd: Through the month just passed, the weather has been very cold and the spring backward. Crops of all kinds are very poor and winter ones are coming in very light. Capt. Merritt has fined us again for non-performance of military duty. And to obtain a final discharge, Buffing, Dodge, and myself went to Cornwell last week to obtain a final discharge from the Surgeon Doctor North. While there I called on your old friend Mrs. Mass who has just returned and lives in the old place. It looks just as it did when we boarded there & the same old school house is there. The place has altered but little, nowise except the old meeting house is torn down and a new one erected nearly in front of the Academy, and quite a number of old folks have died. I spent most of the afternoon there and it brought forcibly to my mind the early associations connected with some of the happiest days I veer spent.

Since my last date Chas Flint and Henry Bird have both "gone down." It is supposed that they were both worth nothing. Indeed it is a general time of failures. But it is something new for Salisbury to have so many is so short a time. I do not expect that all my letters will reach you. But shall continue to write, so that out of a number you may possibly receive some. Do write to me as often as you can. Anything will be interesting from you for you are so far away and engaged in new business that it cannot be otherwise. Your Father's family desire to be remembered to you, with much love.

I must bring this letter to a close, and perhaps you wish it, for I do confess it is a poor one. I trust you will excuse me for I am not well neither have I been for 6 or 7 weeks. I have for the most of that time been troubled with Neuralgia in the face, which if you have not had you cannot judge how much one suffers. Don't forget to let me hear from you soon. May God keep, protect and bless you is the sincere wish of your friend

Myron H. Fish

Letter from M.H. Fish to Graham Lee—Pacific Ocean forwarded by Mr. E. Whittlesey

Riga, Sept. 13, 1843

My dear Roderick,

The present offers too fair an opportunity for sending you a letter to let it pass. Mr. Whittlesey who goes to the Sandwich Islands as a missionary will most likely take this to where taken will reach you. If it ever does is a matter of the future. I suppose that by this time you are busily engaged in your occupation and often imagine you harpooning a "monster" of the deep. Or sailing about in a fine climate with now & then a storm that calls all hands on deck. I was at your Father's a few weeks since & they were all well with the exception of Henry who is somewhat dyspeptic and his general health not very good.

There have been some changes in Salisbury within the year past. Your Father is now erecting a fine house nearly opposite the M. Church and contemplates moving into it this fall. Sometime in the spring Mr. J. Blodget lost by theft $2000 in money and being very credulous with his better half also consulted a <u>Witch</u> from Albany. The Hag finally came to S. to hunt up the money. While she was here she made soma remarkable recitations regarding you. Among which is that you are now on your way home. Lots of the same kind which I do not recollect. I should not have mentioned this but for the reason that we have rec'd only two letters from you since you sailed, up to this time, the first one you wrote your Father and one to myself. We are all very anxious to hear of your welfare and hope you will write soon to let us know of your whereabouts etc. You cannot imagine how much pleasure it affords your family to hear from you and it is a source of no less happiness to me. It is now nearly 9 mos. Since your last came to me and it seems almost an age. Since I wrote you last I have been in New York.

While there wished that you were at E. again. I made a short journey for my health and found upon my return it was somewhat improved but I never expect to be well again until I leave this region of fog and snow. It is now so cold here that I am obliged to keep a fire in the stove to make it anyway comfortable.

Our friend J. Reed's troubles have finally terminated by her having united herself with C. Pratt. While on subjects of this kind our friend the Knight is engaged to his Dulcinea. He leaves so report says this state for Pensy. To commence the glorious uncertainty. I know of no others unless it be Coffing who I rather suspect is getting somewhat tender somewhere north of this. "Este Perpetus" is but little regarded among the young ones now a days. S. Sterling, D. Hamblin, Buce Hollister & E. J. Lee are as usual. Edward I understand thinks of attending Lectures in N. Haven the coming winter. J. G. Mitchell is now sick at his Father's. His physicians fear with the consumption. Business as a general thing is now getting better, that is, most kinds. The iron manufacturing is not, as it was the last down so it will be the last up. All Mothers are improving. No thanks to "Capt. Tyler & Co.". The electioneering campaign for the next Presidential election is about commencing and will equal if not surpass that of '40 of which you no doubt have some recollection. Temperance at present is top in most of the N. England states, New York and some of the middle and western states. The cause has done good no doubt but is now being mingled with religion and politics (mostly with the latter) so that it will soon cease to be the cause it once was.

November 5th Sunday afternoon. When I commenced this last Sept. the idea that it would linger along until this time never entered my mind and it would not have done if Whittlesey by whom I intend sending it had not concluded to remain until nearly the 20th inst. And another reason my brother J. has made us a short visit from Ohio which with my usual business put the idea of your letter from my mind. He (J.) is well located in Ohio and practicing physic among the "Buckeyes", married and seems now to have sown his "wild oats" and settled down in life. From his description of the country and prospects for a young man it is more than probable that your humble servant will take up his abode in that region unless something of material importance occurs within the next 18 mos. The longer I live in such a place as this the more I dislike the inhabitants. I mean the wealthy, aristocratic part of the community. It seems to be impossible for one to rise to other eminence or wealth here who has not the means or assistance of wealthy friends.

Your Father has built the past summer a new house nearly opposite the Methodist Church, which is now nearly completed and will occupy in a week or two. The family are all well and waiting with much impatience to hear from you. Edward Lee has left for N. Haven or New York to attend medical lectures this winter. Hollister is keeping school in his father's district and his brother H. in Sheffield. Mitchell is failing slowly with the consumption and M. Wardwell returned from Hudson a few days since, it is said with the same complaint. Since my last date Henry Bird has died. He failed early in the spring and from that time has been gradually failing until he became finally deranged and died so. One item more of news: Mr. A. Reed has finally moved into the Parsonage and 8th inst. gives a donation party as I have just learned. The winter has set in earlier than usual, it is but three weeks since we had a severe snow storm and even now our ponds are frozen and the ground also. I met our special friend Miss Peet some weeks since at a party, the first time since we gammoned (?) then so finely and took especial pains to salute everyone in the room but her. She was looking quite *** (blotted out by sealing wax) at the time especially about the eyes. Miss Harrison is now in New York, success to her, and among all her flirts may she find one for whom she is fitted and who is fitted to her. By the way it is reported that the jolly old chap Capt. G. Lee is getting very tender in the region of Deacon Smith's and has some foul design of robbing the Deacon of his "Wally". Shades of Fergeson! Only think—D. Lyman is now in Hillsdale and frequently appears here in temperance meetings and to illustrate the effect of drunkenness has an occasional spree.

I have racked my brain to give you all the news which at the best is poor enough. Do write as often as you possibly can, If you could only imagine how much satisfaction and pleasure it affords us to get a letter from you, you would write once a week I am sure. Where I may be when you return I do not know but always remember that I have loved you as man should love man and however distant shall expect to meet once more.

      Believe me truly yours,

          J. Furgeson

Letter form Fanny Lee, November 5, 1843

To Mr. Graham Lee, Ship Nimrod, Capt. Sherman, Pacific Ocean
Care of Samuel P. Damon, Honolulu, Sandwich Islands.

Salisbury, Nov. 5th. 1843

Dear Brother Graham,

You see by my date that it is a little more than a year since you left home. We have received but one letter from you and N. H. F. one. We think we should have received more had it been possible for you to send them. After you left many of our friends came to console and some to wish they had so brave a son and it was but a short time before Father and Mother became somewhat reconciled to your leaving. You will I have no doubt like to know what has been going on at home since you left. During the winter there were several parties, two or three rides which were very much like the Egramont with the exception that they did not stay quite all night. Father and Wm.___ settled not much to the satisfaction of either but it is now through with although it has made us a great deal of trouble. Henry says that he will give you the home news and I must write you what has transpired abroad. Then your friend Fish first. Since you left he has abandoned the society of young persons and confines himself entirely to his business. He adheres strictly to the 17th fundamental rule of the Ferguson Society and I have no doubt you will find him as strong a Fergusonian when you return as when you left; he has not been very well of late--—s the dyspepsia, his brother John spent two weeks in town not long ago. He is a physician practicing in the southern part of Ohio. Mrs. Fish says that he is doing better than her other sons. N.H.F. has written you several letters and is going to improve this opportunity. N.L.P. is well and at home where she expects to remain one or two years although Charley is licensed.

She never has spoken to – and says she never will they have met but once since you left. Hannah is in N. York. She might visit _____ if the attractions there were as great as they were two years ago. She is at her old trade still, not at all discouraged by her failure she has spread her net for James Oer but I do not think she will catch him. He is studying law with Mr. Hubbard and bids fair for a smart talented young man. Coffing is just as you left him pursuing the even tenor of his ways living _____ as it becomes a Fergusonian. G. Mitchell is at home sick with the consumption he has been unwell all summer and this fall his disease has terminated in quick consumption and all hope for his recovery is gone he will probably live but a short time his parents are much afflicted. Elliot remained with Mr. Ball til spring when he left for Sharon where he is now studying medicine. You formed a very strong opinion of him if you expressed your true estimate of his character. His talents were inferior to most of our young men, his boon companion was Daniel Pratt. Cousin Edwards in in New York attending lectures, a friend of Dr. Cicknor procured him a place in the institution and he has them gratuitously. Cousin Harriet came home in June and was married in August to Mr. Smith. He is a fine man and a good preacher. They reside in Erie, Pennsylvania. Sam. Church and Lucious Walton a brother of Mary Jewels are attending lectures in N. Haven. San and Lib after one or two love quarrels have both concluded to have nothing more to do with each other. You recollect that Mary Whittlesey had the misfortune to receive a few hundred dollars from Uncle John. Well, Loring Bartley determined to get both herself and property and his father, mother, brothers, and sisters do all in their power to assist. He waited on her all winter and in the spring offered himself and gave her a week to consider and return an affirmative of course he supposed but (you know there is many a slip betwixt the cup and the lip) during that week L. Walton offered himself. She accepted him and vetoed Lo. Walton is a fine young man. Dr. Ticknor with whom he has studied says he will make an eminent physician and we are all glad Mary has him she is one of the best girls we have. A. Chittenden is really married she has chosen Herman Canfield for her lord, a nephew of Dr. B. Ticknor, he formerly attended the Academy in this town. They were married after an engagement of four weeks and have gone to Ohio to reside. Jane Reed and Liberty were married a month ago at West Point. They returned home and made a wedding party and Charles has gone south to spend the winter. L. Whittlesey spent three months since he was licensed, cod fishing for his health. D. Lyman is anything and everything, a Phrenologist, reformed drunkard temperance lecturer: we have temperance meetings once a week at one of those Dr. Ticknor spoke and after him D. Lyman. It was a thin meeting and the speeches were indifferent, he wrote an account of it for a newspaper said it was a rousing meeting most thinking speeches especially the one given by a young man, and after describing himself in glowing colors he says "Would to god we had more such young men." Lu Holister is intending to teach school this winter in the Dr. Ticknor district. He inquires after you often says you promised to write to him his parents, brothers and sisters are all well. Aunt Mary Scoville is very low with the consumption we are expecting to hear every day that she is not living. Aunt Harriet and family spent the summer in town. Although you are away from country and kin I presume you would like to know "how things go at home." In the political world all is commotion. The next presidential election is beginning to excite the attention of the people. The Whigs are all united in Henry Clay and the Democrats divided the, the South for Calhoun, the North for Van Buren and the West for Johnson. Clay possibly be elected. In the religious world Puseyism is exciting the general attention. It is spreading like wildfire among the Episcopalians while the other denominations are uniting to oppose it. The Temperance cause is spreading gradually and even Salisbury is beginning to "wake up to the subject"-- we have temperance meetings once a week and sometimes oftener. We have lecturers from abroad and at home. The ladies too are getting engaged and are about to form a Martha Washington society and we intend to have our 17th fundamental rule too it is <u>total abstinence or no husbands</u>. Now you need not laugh and make the expression that one of our young men did. "I guess they will get a _____ of waiting on." Rachel thinks we could have the rule and be safe. Julia and myself have been teaching school this summer in this village Julia taught a select school for children and I, a district school of 60 scholars. I have not the least doubt but some of mine "will yet be selected." Mr. Phineas Chapin has lost two more of his children. Ruth the wife of George Sterling and Elizabeth the next to the youngest died within seven hours of each other and were buried the same day. Mr. Chapin has but three children left: Harriet Chapin married a Mr. Granger of Barrington, he is a fine man but not very wealthy. Joseph Bingham was here last winter and made a short visit---he came to ask Uncle Samuel to assist him to some business he was very poor was obliged to borrow overcoat and pantaloons and money before he could come. ----My letter you will perceive is very disconnected I have written it in a great hurry at odd spells I commenced it before moving and now finish it after we have been three days in our new house but I hope you will excuse the errors of style and grammar. I think I have given you all the news that will be very interesting. You need not think you will be soon forgotten all our friends both young and old inquire after you very often. N.L.F. has inquired after you several times I thing she has no ill will towards but her feelings were much hurt _____ is not so much esteemed as formerly several stories have been in circulation with regard to his moral character, but I hardly think they are true we shall always esteem him a friend and treat him as such because he was a friend of yours.

Father wishes me to tell you that we all have that affection and esteem for you that we ever have had. The younger ones think that Graham is the beau ideal of perfection and the older ones do not think very differently. When anything is to be done or any difficulty to be surmounted if Graham was only here then every difficulty would quickly be removed. We think of you both day and night in fact you occupy the uppermost place in all our affections. ---Father and Julia would write but they think there is so much uncertainly whether you will ever get a letter that they conclude not to send any. Julia does not like to have her bright thoughts lost, Mother tells me to give you a great deal of love, Ruth wants to have Graham come, he said he would bring her some shells.

Joseph Whitney a son of Paul Whitney has been out two years whaling was at home summer. He called here and gave us a great deal of information about your business. He cruises in Southern Pacific and thought he would not meet you. He went from Mystic, Conn. He goes out as 2nd mate this voyage. We shall send out letters by Eliphalet Whittlesey who is going to the Sandwich Islands as a missionary. He will leave them in the care of Mr. Damon for you. He is to be married next week to a young lady in Newark, New Jersey. Hezakiah has lost nearly all his property and is going to Michigan to live on a farm he has there. Uncle Gay has lost everything and has got to begin the world anew.

Your Affectionate sister,

Fanny

Appendix D

Henry Lee from Furnace Village
February 15, 1844
Addressed to: Mr. Graham Lee
   Ship Nimrod, Capt. Sherman
   Pacific Ocean
   Care of Trarlon Tricketson (?)
   New Bedford May

Dear Brother,

You will probably think it is strange that you have not heard or heard more from your friends in Salisburry but be assured it is not for want of affection or indifference for considering our circumstances and the improbability of your receiving a letter you will pardon us for the omission. Being layed up by a slight wound in the foot I will endeavor to give you a comprehension of Salisburry news, although this sheet will contain but a short one.

Fanny and I wrote to you by E. Whittlesay who sailed for the Sandwich Islands as a missionary in November last. There are the only letters that we have sent. Father received a letter from you last March soon after I read one from you to Fish. We received another from you the 30th of November, 1843 directed from S. Islands by S. C. Damon and one also from him which furnished us quite a treat. We were much pleased id not flattered with the tone of Mr. Damon's letter. He said he came in contact with you very unexpectedly at Morei and spent a day very pleasantly with you. He is a fine man and will be a great benefit to the sailors of the Pacific. You wrote in your last letter that you should return to the Islands in about four months when you would leave a letter for us this we daily expect. You are now fishing for sperm whale and will return to us we suppose in the fall of '45. I shall watch the sign of your coming, but it would be a sad joke if instead of the steeple we should find it under the eaves! As to family affairs I shall say but little (as you received the account in my former epistle) excepting that we are comfortably situated in our new home opposite Dr. Perries enjoying the pleasure of each other's society with that of Aunt Emily who is living with us through the winter. Fish and Milo are attending academy one preparing for the law the other for a clerk in which business he had some experience last summer. My situation since last spring has been anything but aggregable being obliged to remain at home on account of father's circumstances. I shall probably remain at home a year or two although if I had $1,000.00 to cancel the debts I should leave in the spring. I shall build a hope walk in the spring and endeavor to make it through if friends and fortune smile. Respecting the business of the country – It is undoubtedly improving, the Tariff which was adopted in '42 is still in operation and there are but a few things in the history of our government which have done more for the benefit of the country manufactories which were expiring when you left are now1 part reviving under its genial influence. A little experience has proved that not only the manufacturer but the farmer, mechanic and labourer share equally its protection.

(Feb. 24th I have put off the remainder til after the convention that I may give some account of politics). The influence of the Tariff id plainly seen in S. nearly all the iron works are in operation. A factory is to be built within the village this year by A. H. Holley.

Concerning politics—the present state of the W1hig Party is similar to 1840. Henry Clay the Whig nominee. The Van Buren the opposition. The young Whigs of S. have formed a Clay Club in which G. V. Coffing, Fish and others take an active part. There was a convention at Hartford on the 22 of Feb. which I attended with between 35 and 40 of Salisbury folks. There were between 8 & 12,000 delegates at the convention each town striving to get a splendid banner, which were offered by the Hartford Clay Club to the town that would send the most delegates. Considering circumstances worth $200 Salisbury stood among the first but Bristol took it. Whittlesey who is now living there was appointed to receive it and reply to the President who presented it with a speech. Both were excellent. I came across C.W. Calhoun. He wished himself on the same deck with you. He is going to Norwich to be a head clerk in a store room. I could fill 2 or 3 sheets giving an account of the convention. Sprouts, Coffin, Fish, Sterling, Trot, Burke and myself comprised one load you may imagine the rest. As the success and continuance of the Tariff depends upon the success of the Whig party S. will do her best. The said banner reverts to the town that polls the most votes in the spring. L. Burttett who is teaching school near Hartford joined us as a delegate. His history which is that of disappointed hopes I will leave till you return.

We received a letter from you yesterday dated N. W. Coast, Sept. 14th, '43 which was to us of course of much joy. You give a heart wrenching account of the licentiousness of the sailors, may God help you to overcome that temptation and keep others from going in the path which of all others is the most direct to ruin.

Last evening we were favored with a visit form Cog, Hollister, Edwards, M. H. and Wm. Fish, M. H. read us part of his letter, Edwards had just returned from N. York where he has been attending Lectures.

You expressed a wish that you might find your parents were as you left them. But providence has ordered otherwise and for the best you may expect to find them with more or less of their children in Furnace Village (when you return to S. without any further instruction). You were somewhat homesick when you wrote your record but appeared in good spirits in your last.

Society is similar to what it was in '41 & '42 some additions by younger ones & some growing in and some from out of town particularly a son of Rev, Perry of Sharron who is the Devil's own son yet a scholar and a gentleman. Some have left town and some have married off particularly Mary Lee Whittlesey and Lucius Woolton, brother to Mary E. whose nuptials were celebrated on last Wednesday eve, last winter was celebrated for its number of rich parties and this winter they have the same outline of operation but it has come to a curious state like the locomotive. Some young men got on to much Ale wholie steam and ran off the track.

But enough of this I kept a journal from the time that you left til last summer when Fanny commenced I hope that you have kept one.

Aunt Mary Scoville died a week or two with consumption. I will endeavor to send her obituary with a bundle of papers that will go with this letter.

John S. Mitchel is very low with consumption and you will probably never see him again. I watch with him once a fortnight. I will with pleasure bear your greetings to all your friends and relative except for Elliot who after ingratiating himself into the good graces of Miss Luchael left town and is studying medicine in Sharron.

Before I close this letter I must say something about myself. I have been offered a good business if I would get <u>married</u>! But no—I am going to stay at home, live on sweet cake and pie and enjoy the sweets of celibacy a spell longer.

Your friends all speak of you often especially Miss Peel and Leech. I hope when you return you will receive the attention that they will all be glowing to pay to you.

From your ever affectionate brother Henry

Appendix D

**Added to Henry's letter by
Fanny Lee
February 15, 1844**

Dear Brother:

Henry has left a little spot and says that I may have the pleasure of filling it up. We have just received your letter dated Sept. & if a letter from home gives you but half the pleasure the one from you does us; this poor thing will be read til there is not left a word that has not been read over and over again.

Just for one moment imagine yourself at home. Father comes in and says we have got a letter from Graham. Immediately everything is dropped no matter what we are doing we gather around Father, Mother fixes on her spectacles and leans over to catch and s every word & when he is finished you could not find a dry eye in the whole group, but enough.

We are just as comfortable as can be. Father often says that he has never been more so since he was married & if you was with us I think you would not be able to find a happier family in town. We have a fine house as large as Dr. Berry's & a delightful situation and you must expect to be beset by prayers and entreaties to remain here when you get home.

Henry & I wrote you a letter by E. Whittlesey if you ever get them you will get some news. M. Lee prospers finally has had hard things said about his temperance principles but he has come forward and signed the pledge and if he carries himself straight it may blow over. I wish you was here now such yarns as I could spin you & think you would have your match to beat ones on such marvelous sprees, scrapes, and drunken frolics as old S. has witnessed this winter such you have never seen here.

M. L. F. is a good girl yet; inquires after you every time I see her. Charley is succeeding in business and is much esteemed. H. L. H. the same you left her. She has tried all her charms on James Oer but could not come it so she has turned her attention towards Bob Kellogg. She may succeed as he thinks beauty is the chief thing to be sought for. A. A. Chittenden has married Herman Canfield and gone to Ohio, H. M. Lee married Mr. Smith resides in Pennsylvania. Mary L. Whittlesey to Dr. Lucius Waltom (Mary Jewel's brother) & settled on Uncle John Whittlesey' farm so we have another Dr.

See your male friends are well and inquire of you often – wonder why you do not write. Of your poor dispirited persecuted "female acquaintances" I will say no more for fear it might offend. Your sisters sometimes wonder what strangers will think of us and our mother if our oldest and dearest brother will think and speak so contemptibly of all womankind. Elisha is very much like you, everyone says so, don't despise him Milo is the same queer one. Sarah & Ruth grow tall and handsome. Julia says she shall die with friction of the tongue when you get home. All send love.

Your aff sister Fanny

Remember Julia